SOM

Milestones in Discovery and Invention

Nuclear Physics

HARRY HENDERSON

☑® Facts On File, Inc.

For every person, young or old, who can't stop asking questions
and wondering about the answers.
. . . .

Nuclear Physics

Facts On File, Inc.
11 Penn Plaza
New York NY 10001

Library of Congress Cataloging-in-Publication Data

Henderson, Harry, 1951–
    Nuclear physics / Harry Henderson.
       p.   cm. — (Milestones in discovery and invention)
    Includes bibliographical references and index.
    Summary: Profiles specific physicists, explores their scientific discoveries with a
focus on the process and inspiration involved, and points out how these discoveries
have changed the world forever.
    ISBN 0-8160-3567-9
    1. Nuclear physics—Juvenile literature.   2. Nuclear physicists—Biography—
Juvenile literature. [1. Nuclear physics. 2. Physicists.] I. Title. II. Series.
QC778.5.H46   1998
539.7'092'2—dc21   97-17380

Facts On File books are available at special discounts when purchased in bulk
quantities for businesses, associations, institutions or sales promotions. Please call
our Special Sales Department in New York at 212/967-8800 or 800/322-8755.

Text design by Cathy Rincon
Cover design by Smart Graphics
Illustrations on pages vi, 8, 25, 30, 40, 48, 67, 69, 72, 74, 77, 93, 97, 109, and 120
by Jeremy Eagle

You can find Facts On File on the World Wide Web at http://www.factsonfile.com

This book is printed on acid-free paper.

Printed in the United States of America

MP FOF 10 9 8 7 6 5 4 3 2

Contents

· ·

Acknowledgments

. . . .

I would like to thank the staff of the Niels Bohr Institute, Copenhagen, Denmark; the Niels Bohr Library of the American Institute of Physics; and the public information staffs at the Fermilab and the Lawrence Berkeley Laboratory for their help in finding illustrations for this book.

I would also like to thank the staffs of the Berkeley and Richmond, California, Public Libraries for reference help and interlibrary loan service.

Introduction

. . . .

Nuclear physics is the study of the structure of atoms and the interaction of the particles and energies that make up the very heart of matter. As a science, it is only about a hundred years old, having its beginnings at the very end of the 19th century.

The Atomic Revolution in Physics

The discovery of radioactivity and the inner world of the atom in the late 19th century was unexpected because it did not fit into the ideas of the physics of the time.

Nineteenth-century scientists had made considerable progress in chemistry, the study of how substances combine to form compounds, or break down and produce energy. In 1808, the English chemist John Dalton began a careful study of common chemical reactions. For example, he measured the ways in which nitrogen and oxygen combined into molecules. He summarized his results as follows:

> the elements of oxygen may combine with a certain portion of nitrous gas [nitrogen] or with twice that portion, but with no intermediate quantity. In the former case nitric acid is the result; in the latter nitrous acid . . .

THE PERIODIC TABLE OF ELEMENTS

1																	2
H 1.008																	**He** 4.003

1 ─ atomic number
H
1.008 ─ atomic mass

3	4											5	6	7	8	9	10
Li 6.941	**Be** 9.012											**B** 1.081	**C** 12.01	**N** 14.01	**O** 16.00	**F** 19.00	**Ne** 20.18
11 **Na** 22.99	12 **Mg** 24.31											13 **Al** 26.98	14 **Si** 28.09	15 **P** 30.97	16 **S** 32.07	17 **Cl** 35.45	18 **Ar** 39.95
19 **K** 39.10	20 **Ca** 40.08	21 **Sc** 49.96	22 **Ti** 47.88	23 **V** 50.94	24 **Cr** 52.00	25 **Mn** 54.94	26 **Fe** 55.85	27 **Co** 58.93	28 **Ni** 58.69	29 **Cu** 63.55	30 **Zn** 65.39	31 **Ga** 69.72	32 **Ge** 72.59	33 **As** 74.92	34 **Se** 78.96	35 **Br** 79.90	36 **Kr** 83.80
37 **Rb** 85.47	38 **Sr** 87.62	39 **Y** 88.91	40 **Zr** 91.22	41 **Nb** 92.91	42 **Mo** 95.94	43 **Tc** (98)	44 **Ru** 101.1	45 **Rh** 102.9	46 **Pd** 106.4	47 **Ag** 107.9	48 **Cd** 112.4	49 **In** 114.8	50 **Sn** 118.7	51 **Sb** 121.8	52 **Te** 127.6	53 **I** 126.9	54 **Xe** 131.3
55 **Cs** 132.9	56 **Ba** 137.3	57 **La** 138.9	72 **Hf** 178.5	73 **Ta** 180.9	74 **W** 183.9	75 **Re** 186.2	76 **Os** 190.2	77 **Ir** 192.2	78 **Pt** 195.1	79 **Au** 197.0	80 **Hg** 200.6	81 **Tl** 204.4	82 **Pb** 207.2	83 **Bi** 209.0	84 **Po** (210)	85 **At** (210)	86 **Rn** (222)
87 **Fr** (223)	88 **Ra** (226)	89 **Ac** (227)	104 **Unq** (257)	105 **Unp** (260)	106 **Unh** (263)	107 **Uns** (262)	108 **Uno** (265)	109 **Une** (266)									

Approximate values of atomic masses for radioactive elements are given in parentheses.

lanthanide series	58 **Ce** 140.1	59 **Pr** 140.9	60 **Nd** 144.2	61 **Pm** (147)	62 **Sm** 150.4	63 **Eu** 152.0	64 **Gd** 157.3	65 **Tb** 158.9	66 **Dy** 162.5	67 **Ho** 164.9	68 **Er** 167.3	69 **Tm** 168.9	70 **Yb** 173.0	71 **Lu** 175.0
actinide series	90 **Th** 232.0	91 **Pa** (231)	92 **U** 238.0	93 **Np** (237)	94 **Pu** (242)	95 **Am** (243)	96 **Cm** (247)	97 **Bk** (247)	98 **Cf** (249)	99 **Es** (254)	100 **Fm** (253)	101 **Md** (256)	102 **No** (254)	103 **Lr** (257)

The periodic table arranges the elements by atomic number. The elements in each column have similar chemical properties.

Dalton had observed that chemicals could combine only in certain whole-number proportions. This must mean that there was a certain minimum amount of any chemical element that could not be broken down further. Chemical compounds must be formed by combinations of individual particles, or atoms. Only a whole atom could combine with another whole atom; there were no partial atoms. Further, from the proportions by which elements combined, Dalton was able to calculate the relative weights of their atoms. Dalton made a list that ranked the elements according to their atomic weight.

As chemists continued to experiment with the chemical elements, they noticed that certain groups of them shared common properties, such as their "willingness" to combine with other substances. In the 1860s, the Russian chemist Dmitri Mendeleyev, discovered that the list of elements and their atomic weights could be arranged in table form so that the columns of the table contained elements with similar properties. For example, the alkali metals lithium, sodium, and potassium that shared one column were very chemically active, while the "noble gases" neon and argon found in another column seldom participated in chemical reactions. This arrangement of elements became known as the periodic table because the chemical characteristics were repeated at regular intervals, or periods, down the columns.

Mendeleyev's periodic theory had an important characteristic of good science: predictive value. When the table was arranged so that similar elements were in the same column, gaps sometimes appeared where an element should be, but no such element was known. Mendeleyev boldly predicted that the missing elements would be found someday, and they eventually were.

The End or the Beginning?

By the 1890s, the atomic theories of Dalton and Mendeleyev had run into something of a roadblock. The periodic theory had little explanatory value. That is, it revealed patterns of chemical properties, but neither Mendeleyev nor other chemists of his day had a plausible idea about just what made the atoms of different elements behave differently. Could atoms differ in structure somehow? Even if they did, there seemed to be no way one could examine something as inconceivably tiny as an atom.

Despite such lingering questions, the achievements of 19th-century physics were so impressive that William Thomson, known as Lord Kelvin, an important experimenter in thermodynamics (heat flow), could declare that "Physics has come to an end." Many scientists thought that the 20th century would be merely a time for tying up loose ends. Instead, starting in 1895, the discoveries of X rays, nuclear radiation, and atomic particles, together with Albert Einstein's theory of relative space and time, would revolutionize our understanding of nature at its deepest level.

Physics Enters History

The physicists featured in this book made their discoveries during times of rapid and profound social change. Marie Curie and, later, Lise Meitner entered the field at a time when science, like many other professions, was virtually closed to women. Their lives of achievement would demonstrate that science has no gender and would pave the way for later generations of women.

But in the 1930s, physicists, both men and women, faced another, more deadly challenge. Meitner, together with many other brilliant European physicists, was Jewish. When the Nazis came to power in Germany, Niels Bohr and many others would have to flee their native lands while scientists who stayed behind had to decide whether they would cooperate with Adolf Hitler's government. A bit later, physicists such as Richard Feynman would be called upon to create a terrifying new weapon: the atomic bomb, bringing another set of ethical questions.

When Max von Laue, a prisoner of the Allies after World War II, heard about the atomic bombs, he said:

> When I was young, I wanted to do physics and experience world history. Now that I am old, I can indeed say I have done physics and I have truly experienced world history.

History had truly caught up with this most abstract of the physical sciences. As we approach the end of the 20th century, we can look back on a remarkable history of discoveries about the atom and its hidden particles. But the discovery of natural phenomena also reveals something about ourselves and our human fears and dreams. As Werner Heisenberg once remarked:

> Natural science does not simply describe and explain nature;
> it is part of the interplay between nature and ourselves; it
> describes nature as exposed to our method of questioning.

As you read about the lives and discoveries of the nuclear physicists, you will see how each brought a unique personal style to "doing physics." What do a hearty English gentleman, an intensely driven Polish woman, a philosophical Dane, a cultured Jewish-Austrian woman, and an upstart from New York who loved to play practical jokes have in common? Perhaps it is their curiosity about why nature is held together the way it is, and their perseverance in unraveling each layer of its secrets. Nuclear physics may be about atoms and other particles, but it is also about one of the great adventures of our times.

General Reading for Nuclear Physics

The following books deal with nuclear physics in general or with the work of more than one of the scientists featured in this book.

Apfel, Necia H. *It's All Elementary: From Atoms to the Quantum World of Quarks, Leptons, and Gluons.* New York: Lothrop, Lee & Shepard Books, 1985. Excellent summary of atomic particles for intermediate readers.

Berger, Melvin. *Atoms, Molecules and Quarks.* New York: G.P. Putnam's Sons, 1986. A somewhat shorter and simpler summary for less experienced readers of key developments in understanding the behavior of atoms and their parts.

Bickel, Lennard. *The Deadly Element: The Story of Uranium.* New York: Stein and Day, 1979. Compelling narrative of research into radioactivity; includes accounts of Marie and Pierre Curie, Ernest Rutherford, Niels Bohr, and others.

Boorse, Henry A.; Motz, Lloyd; and Weaver, Jefferson Hane. *The Atomic Scientists: A Biographical History.* New York: John Wiley, 1989. Good survey of the history of nuclear physics. Includes many mini-biographies of important physicists.

Crease, Robert P., and Mann, Charles C. *The Second Creation: Makers of the Revolution in 20th-Century Physics.* New York:

Macmillan, 1986. A fascinating and readable account based on interviews with the pioneers of modern physics.

The Discovery of Radioactivity: The Dawn of the Nuclear Age. Access Excellence Reference Collection. World Wide Web: www.gene.com/ae/AE/AEC/CC/radioactivity.html. Summarizes how the discovery of radioactivity changed scientific concepts of the atom and its structure.

Henderson, Harry and Yount, Lisa. *Twentieth Century Science.* San Diego: Lucent Books, 1997. A useful overview of modern developments in physics, chemistry, biology, medicine, and other sciences.

Ne'eman, Yuval and Kirsh, Yoram. *The Particle Hunters.* Second ed. Cambridge: Cambridge University Press, 1996. A rather textbooklike presentation, but a systematic and detailed overview of particle physics, important physicists, and their discoveries. Includes math that can be followed by advanced high school algebra students.

The New Physics: The Route into the Atomic Age. Edited by Armin Hermann. Bonn: Inter Nationes, 1979. A collection of papers about the key people and places in the story of modern physics. Illustrated.

Radioactivity: Historical Figures. Access Excellence Reference Collection. World Wide Web: www.gene.com/ae/AE/AEC/CC/historical_background.html. The introduction states that "This article will focus on the efforts of four scientists: Wilhelm Conrad Röntgen, Antoine Henri Becquerel, Marie Sklodowska Curie, and Ernest Rutherford. It emphasizes their contributions to the elucidation of radioactivity and the 'key' experiments they performed pertaining to their discoveries."

Segrè, Emilio. *From X-Rays to Quarks: Modern Physicists and Their Discoveries.* San Francisco: W. H. Freeman, 1980. Translation of an Italian work that combines biography and a careful account of discoveries and the development of key concepts.

NOTES

p. v   "the elements of oxygen . . ." Quoted in Boorse, Henry A.; Motz, Lloyd; and Weaver, Jefferson Hane. *The Atomic Scientists: A Biographical History*. New York: John Wiley, 1989, p. 37.

p. vii   "Physics has come to an end." Quoted in Vansittart, Peter. *Voices 1870–1914*. New York: Franklin Watts, 1985, p. 194.

p. viii   "When I was young . . ." Quoted in *The New Physics: The Route into the Atomic Age*. Edited by Armin Hermann. Bonn: Inter Nationes, 1979, p. 105.

p. viii   "Natural science does not simply . . ." Quoted in Mackay, Alan L. *Scientific Quotations: The Harvest of a Quiet Eye*. New York: Crane, Russak, 1977, p. 72.

Key to Icons in Boxed Features

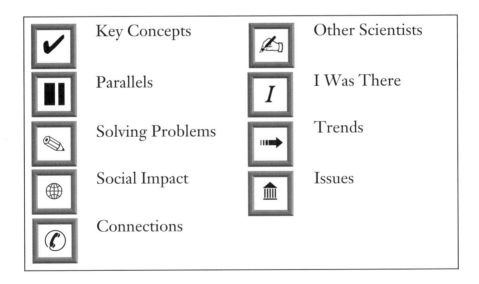

✔	Key Concepts	✍	Other Scientists
❚❚	Parallels	*I*	I Was There
✎	Solving Problems	▫➡	Trends
⊕	Social Impact	🏛	Issues
✆	Connections		

Discovering Radiation

THE CURIES AND THE QUEST FOR RADIUM

Marie Curie's determination helped launch nuclear physics—and scientific careers for women. (Courtesy of the Archives, California Institute of Technology)

1

"I. . . . .
t's radioactive!"

Hearing that phrase brings a shiver of fear to the spine of most people. Nuclear radiation, a kind of energy less tangible than fire or even electricity, seems to combine mystery and a sense of danger. Radiation can both heal and kill. Treatment with radiation in a hospital can cure some cancers, but radiation is also one of the environmental causes of cancer.

Radioactive substances are also associated with the possibility of death and destruction on a vast scale. In August 1945, two atomic bombs destroyed the Japanese cities of Hiroshima and Nagasaki, killing tens of thousands of people in a blinding flash and mutilating and sickening many more. During the years of the cold war (from about 1950 to 1990), the United States and the Soviet Union could have destroyed one another, and much of life on Earth, by pushing the buttons that would launch thousands of missiles tipped with nuclear warheads.

But to the physicists of the 1890s, the discovery of radioactivity brought a different set of feelings: a mixture of perplexity, wonder,

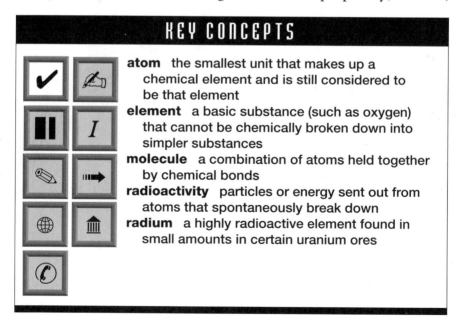

KEY CONCEPTS

atom the smallest unit that makes up a chemical element and is still considered to be that element

element a basic substance (such as oxygen) that cannot be chemically broken down into simpler substances

molecule a combination of atoms held together by chemical bonds

radioactivity particles or energy sent out from atoms that spontaneously break down

radium a highly radioactive element found in small amounts in certain uranium ores

OTHER SCIENTISTS

WILLIAM CROOKES AND CATHODE RAYS

Around the middle of the 19th century scientists began to experiment with electrical spark discharges from high-voltage coils. At first the sparks were sent through the open air. Glassblower and technician Heinrich Geissler then began to make glass tubes from which much of the air had been removed. When electricity was discharged into a Geissler tube, the tube did not make a spark but rather lit up with a steady glow. The glow became known as "cathode rays" because it came from the cathode, or negative electrode, that introduced the electric discharge into the tube.

In 1879, William Crookes, already an accomplished scientist, inventor, businessperson, and editor, wrote a key paper that explained much about cathode rays. Crookes then studied the effect of electrical discharges on a radiometer, a detector that uses a foil strip that moves in response to light or other electromagnetic radiation. He determined that "cathode rays" were not light but rather a stream of charged particles or "projected molecules." He proclaimed that "the phenomena in these exhausted [airless] tubes reveal to physical science a new world." The discovery of X rays 16 years later would prove Crookes to be right.

and curiosity. Atoms, those solid little balls that had proven so useful for understanding chemistry, were not supposed to pop and spark and fire off mysterious rays all by themselves. The study of radiation quickly led to an outpouring of questions about the tiny parts that made up atoms, and the powerful forces that held those parts together.

Much of the first systematic work in understanding radioactive substances was done by an unlikely newcomer to the orderly world of European science. This young but determined researcher came from Poland, a Russian-occupied country on the threshold of Eastern Europe that was often looked down upon as backward by French and Germans. And even more remarkably, the Polish visitor to the great French university at the Sorbonne was a woman named Marie Curie. A woman physicist or chemist was something almost unheard of in 19th-century Europe. Although it would take years before they truly crumbled, the rigid social barriers that had kept women out of science were, like the sparking atoms, starting to break down.

Two Years of Wonder: X rays and Radioactivity

The new physics was born in a quiet but dramatic way. German physicist Wilhelm Conrad Röntgen reported that one day

> I was working with a Crookes [cathode ray] tube covered by a shield of black cardboard. A piece of barium platino-cyanide paper [a material that glows when struck by radiation] lay on the bench there. I had been passing a current through the tube and I noticed a peculiar black line across the paper. . . . The effect was one which could only be produced [ordinarily] . . . by the passage of light. No light could come from the tube, because the shield which covered it was impervious to [could not be penetrated by] any light known. . . . I assumed the effect must have come from the tube, since its character indicated that it could come from nowhere else. I tested it. In a few minutes there was no doubt about it.

Röntgen had discovered a form of energy that could pass through solid objects the way ordinary light passes through a window. He called this new form of radiation X rays, with the X being a symbol in algebra for something unknown. Many other people called them Röntgen rays.

Almost immediately doctors began to use X rays to help them see where bones were broken. For physicists, however, X rays were significant in other ways. They confirmed James Clerk Maxwell's prediction that scientists would someday find electromagnetic waves that had wavelengths much shorter than those of visible light. They suggested that a whole range of invisible energies might be flooding the universe. X rays also gave physicists a new tool with which to probe the structure of matter as well as the effects of energy on individual atoms.

It was natural for scientists to systematically explore the effects of X rays on a variety of substances. One such experiment was carried out in 1896 by Antoine Henri Becquerel. For three generations, the Becquerel family had been studying phosphorescence, or the ability of some substances to continue to glow after having been exposed to light.

Like many other scientists, Becquerel wondered whether Röntgen's X rays might actually be a form of phosphorescence. Working in the same laboratory that had been used by his grandfather and father, Becquerel took a compound of uranium, potassium, and sulfur and exposed it to sunlight while it lay on a photographic plate wrapped in black paper. The paper blocked sunlight, but it would not stop X rays. If the pitchblende was producing X rays in response to the sunlight, the rays would darken the film, a phenomenon that Röntgen had already shown. Indeed, Becquerel found that the film darkened as he expected.

The next few days were cloudy, however. Nevertheless, in one of those lucky moments in the history of science, Becquerel decided to look at the film anyway. He was astonished to find that the film had darkened as much as before. In his report to the French Academy of Sciences Becquerel thus concluded, "There is an emission of rays without apparent cause. The sun has been excluded."

Pitchblende contained the element uranium. Becquerel found that the more uranium a pitchblende sample contained, the more intense the radiation was. He concluded, therefore, that the radiation came from the uranium. In spite of what he had first thought, he also found that the rays produced by uranium compounds were not the same as X rays. This second kind of new rays soon came to be known as "Becquerel rays."

A Unique Partnership

Marie Curie (1867–1934) was born in Warsaw, Poland, as Manya (or Maria) Sklodowska. As a young girl, Marie impressed her parents and teachers with her remarkable memory. At age 16 she graduated from the equivalent of high school with a gold medal for excellence. Marie's father was a teacher of mathematics and physics, so Marie had ample opportunity to become interested in science. But Marie's father lost his savings because of a bad investment. After graduation, Marie had to become an elementary schoolteacher to help support her family. Nevertheless, she was determined to continue her own education.

Marie Sklodowska faced several obstacles, however. Poland was under Russian control, and Russians dominated the education

Pierre Curie had already gained a reputation for his study of magnetism when he met and married Marie Curie. Together, they formed the most famous husband-and-wife team in scientific history. (Courtesy of the Archives, California Institute of Technology)

CONNECTIONS

ELECTRICITY AND CRYSTALS

Piezoelectricity turned out to be quite useful in the 20th century. Since piezoelectric crystals (such as quartz) turned mechanical pressure (such as from sound waves) into electric current, they could be used in microphones and "pick-ups" for record players. A vibrating quartz crystal can also be used to regulate an electric circuit to create a watch that keeps very accurate time or a radio receiver that can be tuned to a precise frequency.

Piezoelectricity also helped researchers learn more about radioactivity. As they worked with radioactive substances, the Curies discovered that "Becquerel rays" electrically charged the air through which they passed. Pierre invented an electroscope that measured the change in shape of a quartz crystal under the influence of electricity. This sensitive instrument enabled the Curies to measure the presence and intensity of radiation coming from a tiny sample of a substance.

system and suppressed Polish culture. Like many bright young adults, Sklodowska became involved in an underground "free university" that defied the Russians by teaching in the Polish language. But career opportunities for a Polish woman in Warsaw were very limited.

In 1891, Sklodowska was able to save just enough money to go to the Sorbonne in Paris for advanced study in science. Because of her poverty, she had a diet of only bread, butter, and tea and lived in a drafty attic room. Despite these hardships, Sklodowska earned advanced degrees in both physical science and mathematics. In 1894, she met Pierre Curie, and they married in 1895.

Pierre Curie (1859–1906) was eight years older than Marie, and he had already made some important discoveries in physics. His

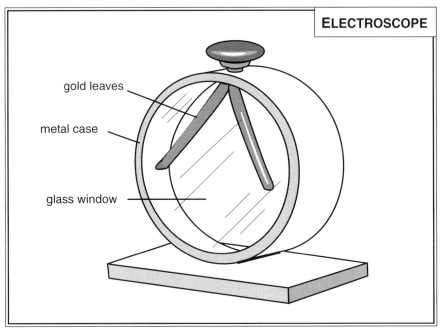

ELECTROSCOPE

gold leaves

metal case

glass window

The electroscope is a simple but sensitive way to detect electrically charged particles such as alpha and beta radiation. When charged particles are present the metal leaves move apart.

most significant work dealt with piezoelectricity, a phenomenon in which certain crystals generated electric current when pressed upon, and in turn, responded to electric current by changing shape.

The Quest for Radium

After their marriage, Pierre and Marie became full partners in scientific work. Marie was looking for a subject for her doctoral thesis. She was excited by Becquerel's discovery of uranium radiation. Was uranium the only chemical element that had this strange property, Marie wondered, or might there be others? Marie decided she would study the radiation systematically, and Pierre helped by creating crude but serviceable laboratory equipment.

Marie soon discovered that a rare element called thorium also produced radiation. In a flash of intuition Marie generalized the concept of radiation, by showing

> the radioactivity of these substances to be decidedly an atomic property. It seems to depend on the presence of atoms of the two elements [uranium and thorium] in question, and is not influenced by any change of physical state or chemical decomposition.

The work of the Curies had changed radiation from a peculiar property of one element, uranium, to a general property that some kinds of atoms apparently possessed. Marie coined the term *radioactivity* to describe the phenomenon.

But Marie found that two uranium ores did not follow the same rules as the others. She extracted all the uranium from one of these ores, called pitchblende (a uranium oxide). But even after all the uranium had been accounted for, the remaining material was four times more radioactive than the uranium itself! In a second stroke of intuition, she realized that there must be a yet undiscovered element hidden in the remainder. It must be a substance that was even more radioactive than uranium.

The unknown element existed in a tiny amount, like a needle in a haystack—but it was a radioactive needle that could be detected by the Curies' instruments. Grinding, sifting, and applying chemicals, they separated out the radioactive material from the rest of the ore. They found a compound of sulfur and bismuth that was highly radioactive. Ordinary bismuth wasn't radioactive, so there must be a small amount of a highly radioactive substance mixed with it. Marie named this substance polonium, after her native country.

But another surprise was waiting. The liquid left over after the bismuth and polonium had been filtered out was still radioactive. The only known element in the liquid was barium, which like bismuth, was nonradioactive. Just as bismuth had a radioactive "chemical cousin," polonium, the barium, too, must have a radioactive relative. Marie later wrote that

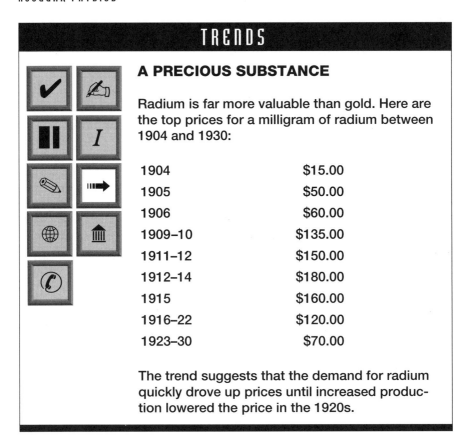

TRENDS

A PRECIOUS SUBSTANCE

Radium is far more valuable than gold. Here are the top prices for a milligram of radium between 1904 and 1930:

1904	$15.00
1905	$50.00
1906	$60.00
1909–10	$135.00
1911–12	$150.00
1912–14	$180.00
1915	$160.00
1916–22	$120.00
1923–30	$70.00

The trend suggests that the demand for radium quickly drove up prices until increased production lowered the price in the 1920s.

after several months more of close work we were able to separate this second new substance. In December, 1898, we could announce the discovery of this new . . . element, to which we gave the name of radium.

So far, however, the Curies had identified radium mainly through indirect means, by concentrating radioactive material. To convince the community of scientists that radium—and polonium—were real, the Curies would have to isolate and purify enough of the substances so that physical and chemical tests could be made. These tests would determine the properties of the new elements, including their atomic weight, and thus their place in Mendeleyev's periodic table. Since radium was easier to extract than polonium, they decided to start with that element.

Chemistry by the Carload

When most people today think of a physics laboratory, they imagine scientists in clean white coats writing at a blackboard—or perhaps a computer screen. The room is filled with sleek, powerful equipment that cleanly delves into the invisible world of the atom. But the task the Curies faced was very different. What they had to deal with was chemistry by the ton, a wagonload of pitchblende at a time, and only a tiny whiff of physics in the form of the elusive radioactivity. Their chemical operations were carried out under primitive and even dangerous conditions. Biographer Rosalynd Pflaum describes Marie Curie's long months of effort like this:

> She was doing the work of a day laborer, keeping the fires going under these [cast-iron cauldrons] while she mixed the poisonous sludge with a steel rod taller than she. There were no exhaust hoods to carry off the noxious fumes emitted by hydrogen sulphide and certain other chemicals, so, in order not to asphyxiate [choke] herself and Pierre, Marie had to continue the distillation process in the open courtyard.
>
> If it snowed, she would freeze; if there was a real downpour, she had no choice but to cart everything back indoors and work with all the windows open. And, of course, when it rained, she also had to rush to move the great jars of precipitates, already produced, from this side to that, to avoid the splash of raindrops trickling down from the cracks overhead. Eventually, as the brews became increasingly concentrated and correspondingly more radioactive and pure, it was essential to keep the laboratory spotlessly clean, an almost impossible job. More than one finished batch, representing weeks of backbreaking labor—pulverizations, crystalizations, precipitations, leachings—was ruined by their woefully inadequate facilities.

In 1903, an exhausted Marie and Pierre Curie learned that they had been awarded the Nobel Prize in physics, together with Henri Becquerel, for the discovery of radioactivity. The husband-and-wife science team became famous overnight, but the fame took time away from their work. Tragedy struck in 1906, when Pierre Curie, crossing a street while lost in thought, was run over and killed by a horse-drawn wagon. The future of radium seemed to be in Marie Curie's hands alone.

In 1911, the prestigious French Academy of Sciences had an opening because of the death of a member. Marie Curie thought that she deserved this ultimate recognition from the French scientific establishment. But despite her Nobel Prize, Marie Curie was not elected to the academy in 1911. Many members thought that women should not participate in the group, and Curie's forceful personality may have rubbed others the wrong way.

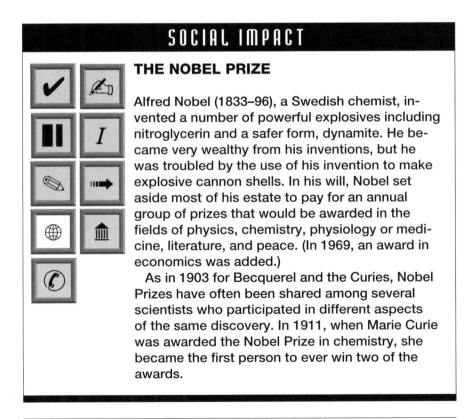

SOCIAL IMPACT

THE NOBEL PRIZE

Alfred Nobel (1833–96), a Swedish chemist, invented a number of powerful explosives including nitroglycerin and a safer form, dynamite. He became very wealthy from his inventions, but he was troubled by the use of his invention to make explosive cannon shells. In his will, Nobel set aside most of his estate to pay for an annual group of prizes that would be awarded in the fields of physics, chemistry, physiology or medicine, literature, and peace. (In 1969, an award in economics was added.)

As in 1903 for Becquerel and the Curies, Nobel Prizes have often been shared among several scientists who participated in different aspects of the same discovery. In 1911, when Marie Curie was awarded the Nobel Prize in chemistry, she became the first person to ever win two of the awards.

Another obstacle came when a rumor arose that Curie and her friend the physicist Paul Langevin had started a romantic affair that threatened his marriage. Modern printing technology had made it possible to publish cheap newspapers that specialized in stories of scandal for popular consumption. To this day no one knows whether the rumors were true. It is clear, however, that as a woman Curie was much more vulnerable to such accusations than a man would have been.

The Radium Institute

Hurt by the distraction and controversy, Curie returned to her work. She began to organize a Radium Institute that would do research in radium as well as provide radium treatment for cancer patients. This effort was interrupted in 1914 by World War I. Looking for a way to serve in the conflict, Curie discovered that despite the widespread use of X rays to diagnose injuries, France was woefully unprepared to use X rays in battlefield hospitals. Curie organized an X-ray service that included

> a touring motor-car, arranged for the transport of a complete radiologic [X-ray diagnosis] apparatus, together with a dynamo [generator] that . . . furnished the electric current necessary for the production of the rays. This car could come at the call of any of the hospitals, large or small, in the surroundings of Paris.

When the war ended in 1918, the Radium Institute went into operation. The one thing it lacked, ironically, was a sufficient supply of radium; there was only enough for medical uses with none left over for research. An American journalist, Marie Mattingly Meloney (nicknamed Missy), came to the rescue. She had decided that Curie's plight would be both a good story for her woman's magazine, *The Delineator*, and a worthy cause in its own

SOCIAL IMPACT

RADIATION TREATMENTS

The discovery of radioactivity provided doctors with one of their most powerful weapons against cancer. Because they are growing quickly, cancer cells are more vulnerable to radiation than normal cells are. If a cancer has not spread too far into the body, it can often be destroyed by a narrowly focused beam of radiation. Today computer-controlled radiation beams make treatment even more precise and less damaging to the normal tissue that surrounds the tumor.

Besides beams, there are other ways radiation can be delivered. Sometimes small capsules containing radioactive material are embedded directly into the tumor. Doctors can also take advantage of natural concentrations of chemicals in the body. For example, the thyroid is a gland in the throat that helps control growth, but it can sometimes become enlarged and overactive. Because the thyroid naturally absorbs iodine, a radioactive version of iodine, when injected into the bloodstream, goes to the thyroid and concentrates its radiation there, shrinking the gland.

right. Although Curie had her fill of sensationalist reporters, Meloney persuaded her to visit America on a publicity tour that raised the $100,000 price of a single gram of radium—a mass equal to about $2/5$ of a dime!

Exhausting work and the effects of decades of exposure to radiation took a terrible toll on Curie's health through the 1920s and until her death in 1934. These problems included cataracts (a clouding of the lens of the eye), burns, sores, and eventually anemia (destruction of blood cells).

For later women scientists such as Lise Meitner, Marie Curie had shown what was possible. According to biographer Robert Reid:

> As a woman scientist she was liberated because she had created the conditions for her own liberation. She had tackled her profession's problems as an equal to all the rest involved; and all the rest happened to be men. She had expected no concessions and none had been made. She had survived because she had made men believe that they were dealing not just with an equal, but with an intensive equal.

The work of Marie and Pierre Curie had shown that radioactivity was a natural atomic property. Two new elements, polonium and radium, had been added to the periodic table. Most important, the Curies had given science a powerful new tool that would soon be used to probe into the very heart of the atom.

SOCIAL IMPACT

RADIUM'S HIDDEN DANGERS

Besides medicine, many other uses were soon found for radium. Some of these uses amounted to scams. Various brands of beauty creams and other cosmetic products claimed to be "radium-enhanced"—fortunately for their users, few of the products were actually radioactive.

Adding a little radium to zinc sulfide and mixing it with paint resulted in a paint that glowed in the dark. This paint could be useful for making watches and instrument dials that could be read at night. But the workers, often women, who applied the radioactive paint, were not warned of its dangers. They often moistened the tips of paintbrushes with their tongues and later contracted cancer.

Chronology of the Curies and the Discovery of Radiation

1808	John Dalton introduces theory that chemical elements are composed of atoms of differing weights
1860s	Dmitri Mendeleyev develops periodic table and predicts the discovery of new elements
1895	Conrad Röntgen discovers X rays
1896	Henri Becquerel discovers the radioactivity of uranium
1898	Marie and Pierre Curie discover radium and polonium
1903	Marie Curie, Pierre Curie, and Henri Becquerel share the Nobel Prize for their discovery of radioactivity
1911	Marie Curie wins a second Nobel Prize, for chemistry
1914	Curie organizes wartime X-ray service
1918	Curie founds Radium Institute in University of Paris
1921	Curie tours the United States to raise money for radium research

Further Reading

Curie, Eve. *Madame Curie* translated by Vincent Sheehan. New York: Pocket Books, 1946. Biography of Marie Curie by her daughter; written in an earlier, more romantic and personal style.

Grady, Sean M. *The Importance of Marie Curie*. San Diego: Lucent Books, 1992. Marie Curie's life and work; for young readers.

Pflaum, Rosalynd. *Grand Obsession: Madame Curie and Her World*. New York: Doubleday, 1989. A richly detailed personal and social history of the Curies and their times.

Reid, Robert. *Marie Curie*. New York: New American Library, 1974. Popular biography.

NOTES

p. 4   "I was working with a Crookes . . ." Quoted in Boorse et al., p. 104.

p. 5   "There is an emission . . ." Quoted in Bickel, Lennard. *The Deadly Element: The Story of Uranium*. New York: Stein and Day, 1979, p. 27.

p. 9   "the radioactivity of these substances . . ." Quoted in Grady, Sean M. *The Importance of Marie Curie*. San Diego: Lucent Books, 1992, p. 34.

p. 10   "after several months . . ." Quoted in Grady, p. 32.

p. 11   "She was doing the work . . ." Quoted in Pflaum, Rosalynd. *Grand Obsession: Madame Curie and Her World*. New York: Doubleday, 1989, p. 77.

p. 13   "a touring motor-car . . ." Quoted in Grady, p. 73.

p. 15   "As a woman scientist . . ." Quoted in Grady, p. 97.

2

Exploring the Atomic World

ERNEST RUTHERFORD'S PIONEERING EXPERIMENTS

Ernest Rutherford's study of radiation led to the discovery of the atomic nucleus and of the transformation of atoms. (Courtesy of American Institute of Physics, Emilio Segrè Visual Archives)

The discovery of radioactive elements by Marie and Pierre Curie had shown that *something* mysterious was going on inside some kinds of atoms. Somehow, atoms of elements like uranium and radium must have a difference in structure that caused them to fire off their mysterious radiation. But if an atom was not the featureless little ball that John Dalton had suggested a century earlier, how could its inner structure be revealed? The first answers would

KEY CONCEPTS

alpha particles positively charged particles ejected from atoms undergoing radioactive decay; they actually consist of a helium nucleus: two protons and two neutrons

beta particles negatively charged particles ejected from atoms during radioactive decay; weaker in force than alpha particles, these turned out to be electrons

electron a negatively charged particle found in the outer parts of atoms

gamma rays a form of very energetic electro-magnetic radiation that accompanies radioactive decay

half-life the time it takes for half of the atoms of a radioactive substance to break down into another substance

ion an atom that has an electric charge. An atom losing a negative electron gains a positive net charge.

nucleus the center of an atom, consisting of one or more protons, neutrons, and a variety of short-lived particles

proton positive particle found in the nucleus of atoms

transmutation the changing of one kind of atom to another as a result of bombardment by a particle

come from the work of Ernest Rutherford, whose experiments would virtually create the science now called nuclear physics.

Like Marie Curie, Ernest Rutherford (1871–1937) was not born in the scientific heart of Europe. Ernest grew up in New Zealand, at the time an obscure corner of Britain's far-flung world empire. His father was a farmer and a wheelwright (a person who makes or fixes wheels on wagons or carriages).

Young Ernest soon showed that he had his father's mechanical ability, and he made himself useful around the shop. The theories of modern physics, like those of philosophers, often seem very remote from everyday life. Rutherford's work, however, would focus on experiments, not theorizing. The practical, hands-on approach of the farmer and mechanic would later help him build the instruments he would need for his experiments. With them, he would hunt the elusive particles inside the atom with the hearty determination of an old-fashioned British big game hunter on safari in Africa.

At the age of 16, Rutherford attended Nelson College, which was more like a modern high school than an institution of higher education. He did well in science and mathematics, but he was also an effective player at rugby-style football. As biographer John Rowland notes:

> It must not be thought, though, that [Rutherford] was a prig or what schoolboys look down upon—a "swot." He took part in all the many activities of the school, and played all the games that were available. But he had remarkable powers of concentration. It was said he could go on reading among the most amazing noise; it was even said that when he was reading it was possible to hit him on the head with a book, and that he would not even notice the blow. But some boys who tried this trick once too often, found sometimes that he did notice it—and then the boy who had hit him found that Ernest had a powerful arm, and was able to use it when necessary.

A small school in rural New Zealand had little equipment for experimental science, however. In 1890, Rutherford won a

I WAS THERE

HAND-CRAFTED INSTRUMENTS

Edward Neville da Costa Andrade, Rutherford's assistant and later biographer, recalls:

> The conditions under which the experimenter did his work were by no means easy. In general, he had to make his own apparatus. P. Lenard, famous for his early work on the electron, told me that he himself made his first induction coil, the apparatus then generally used to produce high potentials [voltages]. This took him several weeks. Sir William Ramsay, the discoverer of the rare gases, made much of his apparatus for handling gases: he was a first-class glass blower.

> Such instruments as could be bought were hand-made and hard to come by. A scientific journal of 1898 suggested: "The possessor of a good induction coil made by our leading instrument-maker should cherish it as the violin player cherishes his Stradivarius or his Guarnerius."

scholarship to Canterbury College, a university in Christchurch, the capital of New Zealand. After earning his degree with honors in mathematics and physics in 1893, Rutherford began his experimental work in a cellar that students used for hanging their coats and hats.

He had become interested in magnetism—in particular, the effects of a rapidly alternating, or changing, electric current on the magnetic properties of iron. This subject might seem to be without practical use. A few years earlier, however, Heinrich Hertz had demonstrated the existence of electromagnetic waves that could travel through the air without wires. These "Hertzian waves," later called radio waves, were generated by discharging stored electricity from one wire coil to another. Rutherford discovered that these waves could make their presence known by causing changes in the magnetism of iron. This research would help lead to the construction of the radio wave detectors that would make Guglielmo Marconi's wireless communication system practical.

Rutherford's work attracted the attention of physicists in Britain, and Rutherford gained a scholarship that would let him travel to Cambridge University for advanced studies. In 1895, Rutherford went to work at the university's Cavendish Laboratory. While cramped and primitive by modern standards, the Cavendish brought together reseachers who would discover many fundamental facts about atomic particles and who would invent basic tools of modern physics.

Beginnings of X-ray Physics

The discovery of X rays by Conrad Röntgen in 1895 had set off a kind of scientific gold rush. Physicists wondered how X rays would interact with the gases they had studied earlier using the less powerful cathode rays. Most of these studies had centered on ionizing a gas—causing its molecules to have an electric charge so they could be manipulated by electrical or magnetic fields. When Rutherford became an assistant to laboratory director John Joseph Thomson and they began their research, they quickly found that "of all the methods by which we can put a gas into a state where it can receive a charge of electricity, none is more remarkable than that of the Röntgen rays."

Running the early X-ray tubes required a deft hand: the gas pressure had to be kept just right—too high, and the electrical discharge would not take place at all; too low, and there wouldn't be enough molecules to sustain it. The electric charge was measured by an electrometer, an extremely delicate device that used a

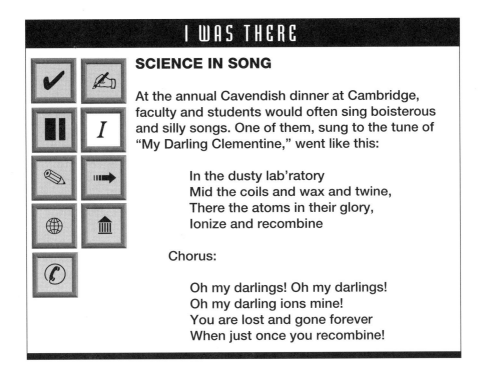

I WAS THERE

SCIENCE IN SONG

At the annual Cavendish dinner at Cambridge, faculty and students would often sing boisterous and silly songs. One of them, sung to the tune of "My Darling Clementine," went like this:

> In the dusty lab'ratory
> Mid the coils and wax and twine,
> There the atoms in their glory,
> Ionize and recombine

Chorus:

> Oh my darlings! Oh my darlings!
> Oh my darling ions mine!
> You are lost and gone forever
> When just once you recombine!

thin strip of gold foil, whose deflection by an electric charge was measured through a microscope. Thomson complained, "This not infrequently refused to hold its charge, and neither prayers nor imprecations [curses] would induce it to do so."

Thomson and Rutherford's paper presented the first precise measurements and mathematical formulas for gas ionization. It showed how the positive and negative ions created by the X rays gradually disappeared as ions with opposite charges were drawn to one another and recombined. Thomson, the senior scientist, was impressed by young Rutherford's careful work.

The Alphabet Particles

Thomson and Rutherford's paper was published in 1897. The year before, French physicist Henri Becquerel had made the

SOLVING PROBLEMS

CATHODE RAYS OR PARTICLES?

John Joseph Thomson became director of the Cavendish Laboratory in 1884 at the age of only 27. While his student Ernest Rutherford worked on X-ray ionization, Thomson again turned his attention to cathode rays. Crookes, inventor of the cathode ray tube, had shown that a magnet bent cathode rays in the same way that a stream of negatively charged particles would be deflected. This suggested that cathode rays were actually a stream of particles, but some physicists still believed that they were dealing with a wave phenomenon. The wave theory, after all, had proven tremendously valuable in the study of electricity, and Maxwell's electromagnetic equations dealt with waves.

Thomson compared the deflection of cathode rays in an electric and a magnetic field. He knew that the effect of an electric field did not depend on the velocity of a charged particle but only on the amount of the charge. A magnetic field, however, creates a current that depends on the velocity of the moving particle. By comparing the magnetic and electrical deflections, Thomson was able to determine that the cathode rays did

astonishing discovery of nuclear radiation in uranium (see Discovering Radiation).

In 1898, Rutherford accepted the position of chairperson of the physics department at McGill University in Montreal, Canada. He wrote to his fiancée, Mary Newton, that "the salary is only 500 pounds (about $2,500) but enough for you and me to start on." When their daughter Eileen was born, he wrote his mother that "it is suggested that I call her 'Ione' after my respect for ions in gases."

indeed consist of tiny particles much smaller thanan atom of hydrogen, the smallest atom. Thomson at first called the particles "corpuscles," but they soon became known as electrons. The discovery of the electron, together with the radiation-spewing atoms that the Curies studied, marked the breakthrough of physics into the subatomic world.

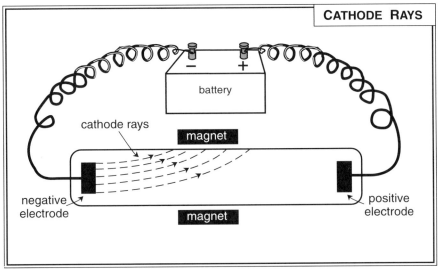

CATHODE RAYS

battery

cathode rays

magnet

negative electrode

positive electrode

magnet

Because cathode rays could be bent by magnetism, it became clear that they were charged particles, not electromagnetic waves.

Rutherford now began to study uranium radiation. He found that his experiments "show that the uranium radiation is complex, and that there are at present at least two distinct types of radiation—one that is very readily absorbed, and the other of a more penetrating character." He called the first kind alpha radiation and the second kind beta radiation, using the first two letters of the Greek alphabet. (In 1900, Paul Villard, a French physicist, discovered the third basic kind of nuclear radiation, called gamma rays, using the third Greek letter. Unlike the alpha and beta radiation that proved to consist of particles, gamma rays are a form of penetrating electromagnetic radiation similar to X rays.)

Studying the radiation from the element thorium, Rutherford found that as thorium broke down, it not only gave off alpha and beta rays, it also produced a mysterious gas or "emanation." The gas, later named radon, was a chemical cousin to gases like helium that were inert, or very reluctant to react with other elements to form compounds.

Rutherford noticed that the amount of radiation produced by a given amount of radon decreased according to a regular pattern. As he reported:

SOLVING PROBLEMS

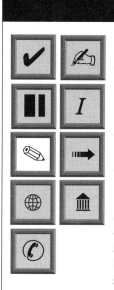

RADIOACTIVE DATING

The fact that radioactive elements decay (break down) at a steady rate makes it possible to determine the age of both rocks and the fossil remains of living things. For example, when a rock is formed, it contains some argon, a gas present in Earth's atmosphere in small quantities. Some of this argon is radioactive and gradually turns into potassium.

Similarly, plants take in carbon dioxide from the atmosphere. Some of the carbon in the carbon dioxide is radioactive. As long as the plant is alive, the level of radioactivity remains constant because the carbon that has broken down is constantly being replaced with new carbon. When the plant dies, however, the carbon can no longer be replaced. The amount of radioactivity steadily declines, and the amount remaining is thus proportional to the age of the plant. This method, called radiocarbon dating, can be used for plants—and animals that eat them—up to about 35,000 years old. Because radioactive argon decays much more slowly, it can be used to date rocks that are hundreds of millions of years old.

In the first 54 seconds the activity is reduced to half value, in twice that time, i.e. in 108 seconds, the activity is reduced to one quarter value, and in 162 seconds to one eighth value, and so on. This rate of decay of the activity of the thorium emanation [radon] is its characteristic feature, and serves as a definite physical method of distinguishing the thorium emanation from that of radium or of actinium, which decay at very different rates.

Rutherford had discovered a universal property of radioactive substances: each substance has a half-life—a specific, unchanging time that it takes for half of the substance to break down and turn into another substance. All radioactive elements eventually break down to a stable "daughter" element, although the process can go through several intermediate stages.

Further, Rutherford's studies of thorium and radon showed that when radioactive elements decayed they turned into other elements. Atoms, far from being unchanging, could turn from one kind into another. "Since therefore," he wrote, "radioactivity is at once an atomic phenomenon and accompanied by chemical changes in which new types of matter are produced, these changes must be occurring within the atom, and the radioactive elements must be undergoing spontaneous transformation."

Rutherford and his assistant Frederick Soddy also attempted to measure the amount of energy involved in the ionization caused by radioactive elements. They discovered that the amount of energy was huge in proportion to the tiny amounts of matter that were breaking down. The last paragraph of their paper made a remarkable prediction:

All these considerations point to the conclusion that the energy latent in the atom must be enormous compared to that rendered free in ordinary chemical change. Now the radio-elements differ in no way from the other elements in their chemical and physical behavior . . . Hence there is no reason to assume that this enormous store of energy is possessed by the radio-elements alone. . . . The maintenance of

TRENDS

SOURCES OF RADIATION

The average person receives the following amounts of radiation (in millirems) per year.

Cosmic Radiation	28
Gamma Rays from Natural Radioelements in the Earth	26
Internal Sources (Radioactive Potassium in the Body)	27
X Rays and Other Medical Tests	38
Total Weapons Fallout	4
Nuclear Plants	<0.01

A millirem is a unit of radiation as absorbed by the human body. A dosage of about 300 rem (300,000 millirem) is fatal to human beings about half the time. These totals vary by location; some people have soil under their houses that releases larger amounts of radon. The total for nuclear plants assumes normal operation, not accidents.

solar energy, for example, no longer presents any fundamental difficulty if the internal energy of the component elements is considered to be available, i.e. if processes of sub-atomic change are going on.

Thus, in 1903, Rutherford essentially predicted the discovery in the 1930s that the sun was powered by atomic energy, plus the idea of nuclear energy that would become the atomic bomb and the nuclear power plant.

Rutherford summarized his discoveries in his book *Radio-activity* in 1904 and won a gold medal from the Royal Society, Britain's most prestigious scientific group. In *Radio-activity* Rutherford summarized the way his researches had led to a new under-standing of the atom:

> The study of the radioactive substances and of the discharge
> of electricity through gases has supplied very strong experi-
> mental evidence in support of the fundamental ideas of the
> atomic theory. It has also indicated that the atom itself is not
> the smallest unit of matter but is a complicated structure
> made up of a number of smaller bodies.

In 1907, Rutherford moved to the University of Manchester, England, which had equipped its laboratories with much better equipment than he had been using at the Cavendish. In 1908, Rutherford was awarded the Nobel Prize in chemistry for his discovery of "the mutability [changeability] of matter and the evolution of the atom."

The Atomic Gun

Rutherford had noticed how the alpha rays sent out by radioactive substances had the ability to penetrate a considerable distance through gas or even thin sheets of foil or glass. Since his earlier research had shown that the alpha "rays" were really positively charged particles, this must mean that much of the space within atoms was actually empty, allowing the particles to pass through.

Rutherford and his assistant Hans Geiger (inventor of the famous radiation counter) created a sort of gun that allowed alpha particles from a radioactive substance to be shot in a beam toward a foil target. By running the apparatus in the dark and straining their eyes, Rutherford and Geiger could see the scintillations, or tiny flashes, where an alpha particle would hit the target, rebound, and flash again as it hit a second target.

Most of the time the alpha particles would bounce off the target at a wide angle. But to their surprise,

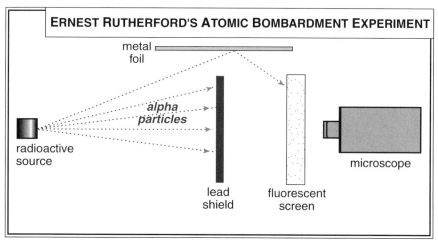

ERNEST RUTHERFORD'S ATOMIC BOMBARDMENT EXPERIMENT

metal foil

alpha particles

radioactive source

lead shield

fluorescent screen

microscope

In Rutherford's atomic bombardment experiment, alpha particles from the source hit the metal foil. Some pass through it, but others are deflected and cause tiny flashes on the fluorescent screen. The microscope is used to measure the deflection angles.

We found that many radiated particles are deflected at staggering angles—some recoil back along the same path they have come. And, considering the enormous energy of the alpha particles, it is like firing a fifteen-inch shell at a sheet of paper and having it thrown back to you!

J. J. Thomson had suggested that atoms were something like a blob of positive electric charge in which the negatively charged electrons were embedded like raisins in a plum pudding. But Rutherford's bombardments had shown that most of the time the alpha particles were passing right through the atoms in the target and being bent only by a small angle. The alpha particles that bounced back, however, had hit something small but hard. Rutherford therefore suggested in 1911 that the atom consisted of a tiny center, or "nucleus," around which the electrons circled, rather like the way the planets in the solar system circle the sun. It soon became clear that the nucleus had a positive charge that held the circling electrons because of the attraction of opposite charges.

Like most scientists, Rutherford had to put most of his work aside for World War I (1914–18). By 1918, however, Rutherford was back in the laboratory, using his favorite ammunition, alpha particles. In 1919, he discovered that when nitrogen gas was bombarded with alpha particles, some hydrogen suddenly showed up. The alpha particles were chipping hydrogen atoms right out of the nitrogen nucleus, leaving an oxygen atom (one unit lower than nitrogen in the periodic table). Rutherford suggested that the nucleii (plural of *nucleus*) of all atoms actually consisted of a number of "hydrogen-sized" particles, with nitrogen having 17 of these, and oxygen 16. Rutherford suggested that these particles be called *protons*, from a Greek word meaning "first ones."

Rutherford's greatest work was now done, but he would continue to take a lively interest in the exciting discoveries of younger physicists during the 1920s and early 1930s. One of these young physicists, Niels Bohr, in a banquet given to honor Rutherford near the end of his life, noted: "If a single word could be used to describe so vigorous and many-sided a personality, it would certainly be 'simplicity.'" And Rutherford called himself a "simple man." In 1932, when he was raised to the nobility, he was asked what title he wanted to use. He chose "Lord Rutherford of Nelson," harking back to the farm town of his birthplace rather than a famous place like Cambridge.

Rutherford had opened up the inner world of the atom to physics much in the way that Galileo had unveiled the solar system. But astronomy had needed an Isaac Newton to fully explain the laws of planetary motion. The atom's Newton would be another good football player, this time one from Denmark.

Chronology of Ernest Rutherford and the Birth of Nuclear Physics

1893   Ernest Rutherford begins to study electromagnetic phenomena

1895   Conrad Röntgen discovers X rays; Rutherford studies ionization effects from X rays

1896   Henri Becquerel discovers radioactivity in uranium

1897   Joseph Thomson discovers the electron

1899   Rutherford discovers that there are two kinds of uranium radiation: positive alpha rays and negative beta rays

1900   Paul Villard discovers a more penetrating form of radiation called gamma rays

1903   Rutherford shows that alpha rays are deflected by electrical and magnetic fields

1904   Rutherford publishes his book *Radio-activity* in which he describes radioactivity as an inherent property of atoms

1908   Rutherford and Hans Geiger invent radiation counter

1911   Rutherford creates a model of the atom that includes a central nucleus

1919   Rutherford demonstrates the transformation of an atom of one element to another

1920   Rutherford suggests the existence of a positive particle (proton) in the nucleus

Further Reading

Andrade, E. N. da C. *Rutherford and the Nature of the Atom*. New York: Doubleday, 1964. Biography of Rutherford that focuses on details of his work.

Eve, A. S. *Rutherford: Being the Life and Letters of the Rt. Hon. Lord Rutherford, O.M.* New York: Macmillan, 1939. Older biography of Rutherford.

Yount, Lisa. *Milestones in Discovery and Invention: Medical Technology*. New York: Facts On File, 1997. Contains a chapter on the work of Conrad Röntgen and the importance of X rays in medicine.

NOTES

p. 20   "It must not be thought . . ." Quoted in Rowland, John. *Ernest Rutherford: Atom Pioneer*. New York: Philosophical Library, 1957, p. 13.

p. 21   "The conditions under which . . ." Andrade, E. N. da C. *Rutherford and the Nature of the Atom*. New York: Doubleday, 1964, p. 8.

p. 21   "The possessor . . ." Quoted in Andrade, p. 9.

p. 22   "of all the methods . . ." Quoted in Andrade, p. 35.

p. 23   "In the dusty lab'ratory . . ." Quoted in Moore, Ruth. *Niels Bohr: The Man, His Science, and the World They Changed*. Cambridge, Mass.: MIT Press, 1985, p. 35.

p. 23   "This not infrequently . . ." Quoted in Andrade, p. 36.

p. 24   "the salary . . ." Quoted in *Encyclopaedia Britannica* [ed.] "Ernest Rutherford" vol. X, p. 106.

p. 25   "show that the uranium radiation . . ." Quoted in Andrade, p. 44.

p. 27   "In the first 54 seconds . . ." Quoted in Andrade, pp. 58–59.

p. 27   "Since therefore, radioactivity . . ." Quoted in Andrade, p. 65.

pp. 27–28 "All these considerations . . ." Quoted in Andrade, p. 72.

p. 29   "The study of the radioactive . . ." Quoted in Wilson, David. *Rutherford: Simple Genius*. Cambridge, Mass.: MIT Press, 1983, p. 196.

p. 30   "We found that many . . ." Quoted in Bickel, p. 46.

p. 31   "If a single word . . ." Quoted in Andrade, p. 16.

3

Mapping the Atom

NIELS BOHR AND THE BIRTH OF QUANTUM PHYSICS

Niels Bohr developed a new picture of the atom based on quantum principles. (Niels Bohr Archive, Copenhagen)

Ernest Rutherford's radioactive bombardments had blazed a trail of discovery into the heart of the atom. By 1911, physicists knew that atoms had a positively charged nucleus containing protons, surrounded by a swarm of tinier negatively charged electrons. The focus of physics now shifted to the explanation of the behavior and arrangement of the electrons. Throughout the 1920s, Danish physicist Niels Bohr would work with Erwin Schrödinger, Werner Heisenberg, Wolfgang Pauli, and other important physicists to develop a new way of looking at the electron and indeed, at reality itself.

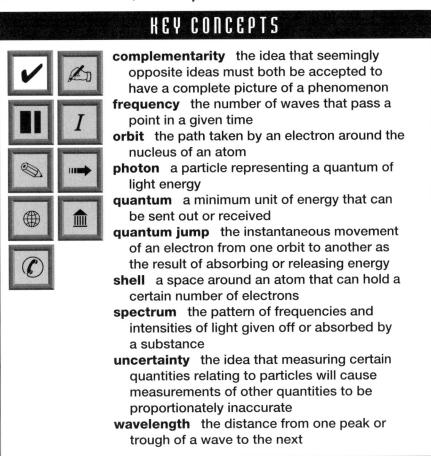

KEY CONCEPTS

complementarity the idea that seemingly opposite ideas must both be accepted to have a complete picture of a phenomenon

frequency the number of waves that pass a point in a given time

orbit the path taken by an electron around the nucleus of an atom

photon a particle representing a quantum of light energy

quantum a minimum unit of energy that can be sent out or received

quantum jump the instantaneous movement of an electron from one orbit to another as the result of absorbing or releasing energy

shell a space around an atom that can hold a certain number of electrons

spectrum the pattern of frequencies and intensities of light given off or absorbed by a substance

uncertainty the idea that measuring certain quantities relating to particles will cause measurements of other quantities to be proportionately inaccurate

wavelength the distance from one peak or trough of a wave to the next

Niels Bohr (1885–1962) was born in Copenhagen, Denmark. Niels's father was a university professor of physiology (the study of life processes), and his mother was also well educated. The Bohr home was frequently filled with the sounds of stimulating intellectual conversation. Perhaps because of this, Niels would develop a lifelong interest in philosophy and literature that would help him overcome the sometimes narrow limitations of scientists who had a less liberal and diverse education.

Young Niels was a good though not outstanding student, but his interest in physical science soon blossomed. Biographer Niels Blaedel notes:

> [Bohr's] brilliance demonstrated itself as a natural and simple thing, without special effort. His mind worked quickly, and even from his schooldays there are accounts of how his thoughts ran faster than his ability to use the eraser when he was standing at the blackboard. He would wipe with both his hands and arms as new ideas rapidly came to him, and neither he nor the blackboard looked too good afterwards.

Young Niels did not spend all his time studying, however. He was an excellent soccer player, almost as good as his brother Harald, who earned a place on the Danish Olympic team.

In 1903, Bohr entered the University of Copenhagen, where he won his first scientific award in 1909, a Gold Medal from the Royal Danish Academy of Sciences and Letters for his study of surface tension—the delicate but surprisingly strong force that creates a kind of "skin" on a liquid. For his master's degree and doctoral degree (in 1911), Bohr did research on the possible arrangements of electrons on the surface of atoms of metals.

Quantum Leap

In 1913, Bohr published his answer to the question of why the arrangement of electrons around the atom's nucleus that Rutherford had suggested was possible. Bohr needed to find something

I WAS THERE

CAUSE AND EFFECT?

Sometimes Bohr relaxed by going with some other physicists to see a movie—often an American western or gangster film. After one such experience, Bohr made a wry comment:

That the villain runs off with the pretty girl is logical; it happens all the time. That the bridge collapses under their wagon is unlikely, but I am willing to accept it. That the heroine continues to hang over the precipice is more than improbable, but I will go along with that. I am even willing to accept that Tom Mix comes riding past at that very instant, but that at the same time there should also be a person with a movie camera recording the entire story— that is more than I am willing to believe.

that would prevent the electrons from radiating energy and falling back into the nucleus, as would be expected under Maxwell's theories of electromagnetism. He found the key in the idea of the quantum—a fixed unit of energy.

Around 1900, German physicist Max Planck had studied how a "black body" radiates energy. A black body is an object that absorbs all the light that hits it, appearing black because it reflects no light. It is a theoretical object that doesn't actually exist in nature, but is useful for studying how energy is radiated.

As a body is heated, it sends out light. As the temperature of the body increases, the light coming from it not only contains more energy, its wavelength grows shorter. This is why a piece of metal, when heated, glows first red, then yellow, and finally "white hot." Red light has the longest wavelength, yellow shorter, and blue-white shortest.

Earlier in the 19th century, scientists had discovered that the light coming from a glowing material was actually a mixture of wavelengths that could be separated out into a spectrum. (A rainbow is a spectrum of colors of visible light.) The different colors represent different fractions of the total energy being emitted, or sent out, by the object. In attempting to apply Maxwell's laws of electromagnetic radiation to find a formula that related temperature to this energy distribution, Planck found that any formula he tried began to give wrong results as one went toward the longer wavelengths.

The "classical" physics of Newton (for moving bodies) and Maxwell (for vibrating electromagnetic waves) both assumed that action was continuous. That is, if one accelerates an object or increases the energy used to vibrate a wave, the resulting movement will smoothly increase according to the appropriate formula. Planck saw that the breakdown of the black body radiation formula must mean that there is a point where the action is not continuous, but comes in tiny "spurts." Planck called one of these spurts a *quantum*, from the Latin word for "how much?" He discovered that for a given frequency of radiation, that radiation can be sent out (or absorbed) only in a "spurt" (or package) whose size is equal to a constant called h (Planck's constant) times the frequency. h is extremely small, being equal to 6.6×10^{-34} units in terms of energy over time, that is, "work" or "action." (10^{-34} means a decimal point followed by 33 zeroes and a one.) Because this unit is so small, the effects of the quantum nature of energy mean very little when dealing with the movement of objects like stars or planets, or even the familiar objects of daily life like cars or basketballs.

But atoms and the particles that make them up are so small that quantum effects not only matter, they often take over. In 1905, Albert Einstein, the German physicist who developed the revolutionary theory of relativity, was studying the photoelectric effect, where ultraviolet light knocked electrons loose from a metal surface. If the light consisted of the "classical" waves of Maxwell, the velocity with which the electrons were kicked loose should steadily increase with the intensity (amount) of the incoming light. But it didn't work that way; it was only the wavelength of the incoming light that determined the electron's energy level.

This meant that the light must be acting like a stream of particles, each with a particular energy level corresponding to its wavelength, as in the black body radiation. Einstein reported: "It appears to me, in fact, that the observations . . . can be understood better on the assumption that the energy in light is distributed discontinuously in space." These discontinuous "bits" of light energy were none other than Planck's quanta, conceived of as particles that Einstein called *photons.* A certain amount of energy had to be applied all at once to knock an electron loose, so this happened only when a light particle had a short enough wavelength to represent a quantum with a high enough energy to do the job. Any number of less energetic photons wouldn't do the job, which is why intensity didn't matter.

Bohr's Quantum Atom

Drawing on what Planck and Einstein had learned, Bohr realized that the idea of the quantum offered a way to keep the electrons in an atom from spiraling in and falling into the atom's nucleus. He said that an electron must always have an angular momentum (basically the energy of orbit) that was equivalent to an integer (a whole number) times *h* (Planck's constant) divided by 2π. (For example, $h/2\pi$ or $2h/2\pi$. π or "pi," as you probably know, is a specific number related to the dimensions of a circle, or in this case, a circular orbit.) Because the energy determines the location of the orbit—as with a satellite circling the Earth—this in turn meant that the electron could only enter a limited number of fixed orbital paths. Like a car on a freeway, it had to be in one "lane" or another, not straddling the line.

Unlike a car, however, the electron does not smoothly change to a faster or slower lane. When light, heat, or some other kind of energy hits an atom, it can cause an electron to instantly jump from an orbit nearer the nucleus to one farther away. The orbit it jumps to depends on the value of the quantum of the incoming

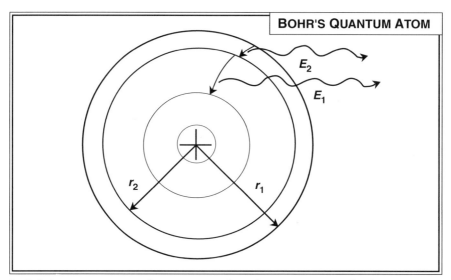

BOHR'S QUANTUM ATOM

In Bohr's quantum atom, electrons can make only certain jumps, absorbing or releasing fixed amounts of energy. Here the orbital jumps r1 and r2 absorb energies e1 and e2.

energy. Similarly, an electron in an outer orbit can jump back into an inner one, sending out a quantum of energy. The motion is instant and abrupt because a whole quantum of energy must be used at one time.

Bohr's quantum theory solved another physics puzzle: what determines the spacing in the lines in the spectrum of a substance? Back in 1885, John Jacob Balmer had discovered a formula that allowed him to calculate the spacings of several series of lines in the hydrogen spectrum on the basis of the squares of integer (whole number) values. Balmer's work may have given Bohr a clue because the presence of integers suggested a quantum in the underlying energy in somewhat the way Dalton's integer proportions in chemical reactions had suggested the existence of a minimum unit—one atom.

Once Bohr developed his quantum theory, he was able to calculate from it the spacing of the lines that Balmer had found. He showed that each spectral line represented a transition from one orbit to another. There were other complications when atoms less simple than hydrogen were involved, but Bohr had taken the essential first step to relating the structure of atoms to the energy they gave out.

The fixed electron orbits of the Bohr atom also offered an explanation for the chemical properties of the elements as shown in the periodic table. In 1925, Wolfgang Pauli, by assigning "quantum numbers" to characteristics of an electron such as its

A modern spectroscope can reveal many details of both atomic and molecular structure. (Courtesy of the Archives, California Institute of Technology)

momentum and spin, showed that no two electrons could occupy the same position in the atom. This became known as the "Pauli Exclusion Principle."

Indeed, the atom turned out to be arranged into orbit "shells" that allowed only a certain number of electrons: 2, 8, 18, and 32 for the first four orbit levels. Each shell is broken down into up to four subshells with slightly different energy levels, which also have limits of 2, 6, 10, and 14.

Bohr's work showed that the most important characteristic of an element is its atomic number—the total number of protons in the nucleus, which normally are electrically balanced by the same number of electrons. Oxygen, for example, normally has an atomic number of 8. Its eight electrons are divided into two in the first shell, and six in the second shell (which in turn is subdivided into subshells of two and four).

Atoms have a tendency to fill their outer shell to capacity. Because the second shell in oxygen can actually hold a maximum of eight rather than six electrons, oxygen has a tendency to bond with two more electrons in order to fill its second shell. Each neutral hydrogen atom has a single electron. If conditions are right, an oxygen atom joins with two hydrogen atoms and shares their electrons. This means that in effect the oxygen atom now has eight outer electrons and the hydrogen, by sharing with an electron from the oxygen atom, fills its single shell with two electrons. The result, is H_2O, or water. An atom of helium, however, already has a full shell with two electrons. As a result, helium does not combine easily with other atoms to form compounds.

While the details are more complex than given here, it is easy to see that Bohr and Pauli had connected physics and chemistry at the atomic level.

Wave or Particle?

While Bohr's theory worked with the simple hydrogen atom, it could not accurately predict the spectra of the more complicated atoms. Physicists now sought a systematic way to calculate the actual paths and energies of electrons in atoms. Two different approaches were tried. In 1923, the French physicist Louis de Broglie suggested that just as Einstein had shown that light, which had been thought of as a wave, has particle properties, perhaps particles also had wave properties. He applied wave formulas to particles such as electrons, so that faster moving particles had shorter wavelengths.

Experimenters then found some startling evidence about the wavelike behavior of electrons. Suppose there is a barrier with two slits in it, placed close enough so that an electron shot toward it has an equal chance of going through either slit. On the other side of the barrier, place a photographic plate that will record where the electrons strike it.

When a beam of electrons is sent toward the barrier, the electrons make the slits act like the sources of two separate waves that interfere with each other and make a characteristic pattern on the film. But suppose one sends not a beam of electrons but just individual electrons, one at a time? After enough electrons have been fired at the barrier, the very same wave interference patterns appear on the film. In other words, even when an electron is being treated just like an individual particle, the place it ends up is determined by its "wavelike" nature!

De Broglie also suggested how Bohr's integer quantum numbers could be explained by wave motion: if the electron were a kind of wave wrapped around its orbit, the wave had to have an integer (whole) number of wavelengths or it would overlap and interfere with itself and be obliterated.

In 1925, German physicist Werner Heisenberg pioneered a different approach to quantum theory. Instead of trying to visualize how the electron actually moved, he and his colleagues compiled sets of numbers that represented various characteristics of the electrons such as energy, momentum, and spin. When these numbers were arranged in squares called matrices and certain mathematical operations performed, the physicist could predict other quantities that could later be confirmed by experiment.

OTHER SCIENTISTS

ERWIN SCHRÖDINGER AND HIS CAT

Erwin Schrödinger was born in Vienna, Austria, in 1887. Like Bohr, he grew up in a home where intellectual discussion was commonplace. He became as fascinated with languages, literature, and poetry as with physics.

Schrödinger's application of wave principles to the motion of the electron came from his unhappiness with Bohr's quantum jumps. He believed he could find an equation that would make the jumps go away in favor of a smooth transition from one level of vibration to another. His wave equation was a masterpiece for which he shared the 1933 Nobel Prize in physics.

To explain how whether one sees waves or particles depends on the observer, Schrödinger came up with his famous cat experiment. This "thought experiment" consisted of a box that contained a radioactive atom, a detector, a poison gas dispenser, and an unfortunate feline. The apparatus was set up so that if the atom decayed and sent out radiation, the detector would trigger the release of the gas and the cat would die. But until the box is opened, the outside observer has no way of knowing whether the atom has decayed, and thus, whether the cat was alive or dead. In a sense, both a live cat and a dead cat exist together until the observation is made, and then there is only one cat, either alive or dead.

Similarly, Schrödinger said, the act of making an observation of the electron "collapses" its "wave function" with its many possibilities into one state and reveals some information about the particle.

In the end, Schrödinger was unable to get rid of quantum jumps or quantum uncertainty, but his wave equation remains indispensable, doing for the electron what Maxwell had done for electromagnetism.

In 1927, Austrian physicist Erwin Schrödinger expanded on de Broglie's theories. De Broglie and Schrödinger thought of the waves as representing a real distribution of electric charge in space. This was troubling because the calculated amounts turned out to include what mathematicians call imaginary numbers. It was hard to visualize what actual physical reality corresponded to these numbers. Further, how could the electron as a tiny package of waves avoid eventually spreading out and vanishing as waves in a pond do?

German physicist Max Born then thought of a different way to look at the electron. He said that the wave was not a physical vibration of energy but rather expressed the chance that the electron could be found in a particular location. This is something like taking two dice and marking boxes numbered 2 to 12 on a piece of paper. If you roll the dice 100 times and mark an X in the square that corresponds to the total of each roll, you create a picture of the probability distribution. Before you roll the dice any given time, you can't know what numbers will come up, but you know that some results (such as 7) are more probable than others (such as 2). Similarly, physicists began to speak of the probability of a particle being in a particular location or having certain characteristics.

Bohr versus Einstein

The fifth Physical Conference of the Solvay Institute in Belgium in October 1927, gave physicists an opportunity to thrash out the exciting but troubling conclusions that Bohr, Schrödinger, and Heisenberg had reached about the paradoxical quantum world.

Quantum mechanics was one of the two great revolutions in 20th-century physics. The other, relativity, was the product of the mind of Albert Einstein, perhaps the greatest physicist of all time. Einstein's theories of relativity, while revolutionary in their conclusions about space, time, and gravity, were classical in their methods. As with Newton's theories, given accurate data, they produced exact results.

Einstein had played an important role in the early development of the quantum idea with his 1905 paper on the photoelectric effect discussed earlier. But Einstein had become deeply troubled by the direction physics seemed to have taken with the theories of Bohr and his colleagues. He insisted "that the essentially statistical character of quantum theory is solely to be ascribed [attributed] to the fact that the theory operates with an incomplete description of physical systems." In other words, if physicists had a better theory and better instruments, the "uncertainty" would melt away.

To try to prove his point to the assembled physicists, Einstein set up one of his famous "thought experiments." Suppose, he said, one sets up a barrier with a single slit in it and puts a photographic plate some distance beyond it. If a single electron or photon went through the slit, Einstein agreed that one could not predict exactly where the electron or photon would hit the photographic plate. One could calculate only probabilities about where it would be found.

Once the particle actually hit, though, that obviously indicated the places where it had *not* hit. Using this data and controlling the electron's momentum and how it transferred energy to the light particle used to observe it, one should be able to work backward and figure out an equation that would predict where an electron with given characteristics would end up.

Bohr objected that if one carefully made the light used to see the electron weak enough to not affect the position of the electron, the light would not be strong enough to let one "see" the electron and fix its position in space.

Einstein then revised his experiment so that a second barrier with two slits was placed between the first slit and the photographic plate. Since the electron was really a particle, Einstein said it could go through one slit or the other, but not both. Again, working back from where the electron was recorded to have hit, it should be possible to determine which slit it had gone through—which meant that contrary to quantum mechanics, the electron's exact position at a particular moment of time would be known. But Bohr again showed that the very attempt to set up the second barrier would result in interference, and thus uncertainty.

In the 1930 Solvay conference, Einstein made one more attempt to refute quantum mechanics. Suppose, he said, one had a box with a source of radiation, a hole, a shutter, and a clockwork mechanism that could open the shutter at a specified time. The box would be put on a spring balance that would show its weight. The shutter would be opened at a precise time, allowing just one photon (light quantum) to emerge. Since Einstein had shown that mass and energy were equivalent, the loss of that photon would result in the scale showing a lower weight for the box. Comparing the "before" and "after" weights would give the mass—and thus the exact energy—of the photon. But determining exact location and energy at the same time was forbidden by quantum mechanics.

Bohr and his colleagues huddled through the night, trying to come up with an answer to Einstein. The next day, Bohr gave his reply. The problem, he said, turns out to be the clock. Einstein himself had shown in his work on relativity that the time indicated by a moving clock in a gravitational field was affected by its movement. In the instant the box opened and the scale moved, the clock's rate of timekeeping would change by an indeterminate amount. Without being able to tell the exact time, the experiment could not give an exact result. It seemed that whatever Einstein tried, uncertainty, like a pop-up toy, would raise its head again.

The friendship between Einstein and Bohr was not affected by their arguments, however. "In spite of all divergences of approach and opinion," Bohr noted, "a most harmonious spirit animated the discussions." When Einstein teased Bohr by asking, "Do you really believe God resorts to dice-playing?" Bohr gently replied, "Don't you think caution is needed in ascribing attributes to Providence [God] in ordinary language?"

From Uncertainty to Complementarity

De Broglie and Schrödinger had shown that it was useful to consider the electron to be a wave. German physicist Werner Heisenberg and his colleagues, on the other hand, looked at the electron as a particle with a number of definite numeric values that described such characteristics as energy, momentum, and

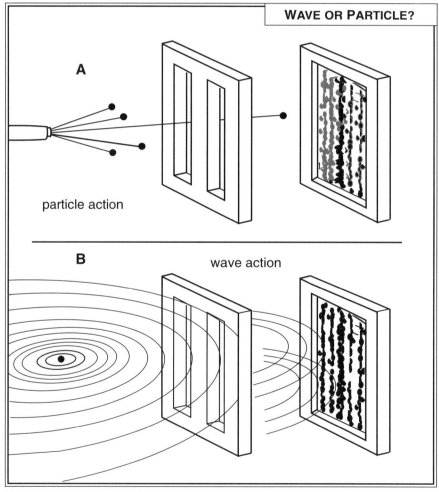

Although an electron can be viewed as a particle (A), it produces the same interference pattern as a wave (B).

spin. Physicists were uncomfortable with the idea that an electron might act like a particle one moment but like a wave the next.

Heisenberg then had an insight that led to a new approach to the problem. In Newton's classical world, the scientist does not have any effect on what he or she observes. When you look at a star or even a butterfly, the light that reflects off its surface and gives your eyes information about the object's location or speed has no effect on the object itself. That is because the object is far larger than the particles of light being used to observe it.

But this is not the case when what you are observing is the size of an electron. If you want to shoot a light particle (photon) at an electron in order to determine its location, you have to use light with an extremely short wavelength so that the waves can measure the tiny electron. But a short wavelength means that the light quantum, or photon, has correspondingly high energy. This high energy photon will give energy to the electron and change its momentum at the instant you measure its location.

Similarly, if you want to measure the electron's momentum, you have to use a photon with a long wavelength (and thus low energy) so as to hit the electron as gently as possible and not change the energy you are trying to measure. But now, while you'll have a good idea of the electron's momentum, you will have a correspondingly less accurate idea of its position because of the longer spacing between the waves.

Planck's constant, that tiny number that is the minimum amount of action that can take place, also turned out to be the limit on accuracy of measurements in the world of the electron. Heisenberg's equation said that the product of the two uncertainties (position and momentum) will always be at least $h/2\pi$.

Like the particle/wave problem, uncertainty seemed threatening to many physicists. After all, precise and exact answers had always been the hallmark of science. But Niels Bohr expressed a different attitude that helped people accept this situation.

While Bohr had certainly participated in the debates between Schrödinger and Heisenberg on waves, particles, and uncertainty during the mid-1920s, he had stayed in the background. "Bohr's role became less that of an initiator in the progress of quantum

OTHER SCIENTISTS

WERNER HEISENBERG (1901–1976)

Werner Heisenberg was born in Würzburg, Germany. His father was a university professor and no doubt encouraged his son to develop academic interests. In high school Heisenberg studied the history of ancient Greek science, and his lifelong interest in philosophy and the history of science prepared him well to deal with the troublesome concept of uncertainty. Young Heisenberg also showed considerable talent in music, as well as the vigorous activities of tennis and mountain-climbing.

Heisenberg's work on quantum mechanics and the uncertainty principle led to a Nobel Prize in physics in 1932. When the Nazis came to power, Heisenberg stayed in Germany. He was strongly opposed to Adolf Hitler, but said later that he thought someone had to try to preserve a bit of culture from the "barbarians." Heisenberg did work on the German nuclear energy program, which lagged far behind that of the Americans and British. He apparently showed Niels Bohr a diagram of a proposed nuclear reactor. When Bohr came to America and drew this picture, it alarmed American scientists because it suggested that Germany had progressed further in nuclear research than it actually had.

physics and more of a support, a mentor, and a penetrating critic of those who were leading the way."

In 1927, Bohr was back on center stage, however, ready to bring a more philosophical viewpoint. He suggested that the seemingly opposite ways of looking at the electron are *complementary* to each other. The more one sets up apparatus to look for a wave, the less like a particle the electron will appear to be. But the more one arranges to deal with a particle, the less wavelike behavior one will find. It is not a matter of particle *or* wave, but rather that particle

SOCIAL IMPACT

ACCEPTING SOME IGNORANCE BRINGS KNOWLEDGE

To the classical Newtonian scientist, the idea of uncertainty was unacceptable because it was a confession of failure. But to Bohr, accepting the uncertainty, the fact that some things cannot be known, was the key to a new kind of knowledge:

> Just the renunciation [giving up of knowledge] forms the necessary condition for an unambiguous definition of the energy of the atom. We must consider the very renunciation as an essential advance in our understanding.

The new physics had a number of pairs of opposite ideas. The electron, for example, seemed to be both a wave and a particle. The position and velocity of an electron were intertwined so that finding out one left one that much more ignorant about the other. Bohr called the idea that two opposite ideas should be held together in one's mind *complementarity*.

Bohr would later try to apply the principle of complementarity to other areas of life. For example, when the long shadow of the Nazis and their belief in racial superiority were starting to spread over Europe, Bohr was asked to address a congress on anthropology and ethnology (the study of human cultures). Bohr said:

> The main obstacle to an unprejudiced attitude toward the relation between various human cultures is the deep-rooted difference of the traditional backgrounds on which cultural harmony is based and which exclude any

(continued)

SOCIAL IMPACT

(continued)

simple comparison between such cultures. In this above all, the viewpoint of complementarity offers itself as a means of coping with the situation . . . we can truly say that different human cultures are complementary to each other.

and wave together are necessary for grasping the full picture. Bohr was saying that the union of seeming opposites lay at the heart of nature and should be embraced as a whole. Gradually, most physicists learned to live with this reality and moved on.

Bohr's pioneering work and the new quantum theories of the 1920s had made it possible to understand the electron. In the 1930s, the focus would shift to the nucleus of the atom and the incredible energies hidden within it.

Chronology of Niels Bohr and Quantum Mechanics

1900   Max Planck uses quantum idea to explain radiation

1905   Albert Einstein uses quantum theory to explain photo-electric effect

1907   Niels Bohr does his first research, on surface tension in liquids

1911   Bohr completes doctoral dissertation on electron theory of metals

1912   Bohr goes to Cambridge, meets J. J. Thomson and Ernest Rutherford. Rutherford publishes ideas about nuclear atom.

1913   Bohr publishes quantum theory of atomic structure

1917   Bohr begins to establish Institute for Theoretical Physics in Copenhagen, Denmark

1922   Bohr receives Nobel Prize in physics for his work on atomic structure

1923   Louis de Broglie uses wave theory to describe behavior of electron

1924   Wolfgang Pauli explains quantum numbers and exclusion principle

1925   Werner Heisenberg, Max Born, and Pascual Jordan formulate particle-based quantum matrix mechanics

1926   Erwin Schrödinger develops alternative wave approach to quantum mechanics

1927   Werner Heisenberg formulates uncertainty principle; Bohr introduces concept of complementarity

Further Reading

Blaedel, Niels. *Harmony and Unity: The Life of Niels Bohr*. Madison, Wisc.: Science Tech Publishers, 1988. Biography that mainly focuses on the scientific work.

Moore, Ruth. *Niels Bohr: The Man, His Science, and the World They Changed*. Cambridge, Mass.: MIT Press, 1985. Life and work of Bohr.

Niels Bohr: A Centenary Volume. Edited by A. P. French and P. J. Kennedy. Cambridge, Mass.: Harvard University Press, 1985. A collection of papers about Bohr's life and work.

Spangenburg, Ray, and Moser, Diane K. *Niels Bohr: Gentle Genius of Denmark*. New York: Facts On File, 1995. A young adult biography.

NOTES

p. 36   "[Bohr's] brilliance . . ." Quoted in Blaedel, Niels. *Harmony and Unity: The Life of Niels Bohr*. Madison, Wisc.: Science Tech Publishers, 1988, p. 15.

p. 37   "That the villain runs off . . ." Quoted in Blaedel, p. 129.

p. 39   "It appears to me . . ." Quoted in Boorse, p. 143.

p. 46   " . . . that the essentially statistical . . ." Quoted in Moore, p. 166.

p. 47   "Do you really believe . . ." Quoted in Moore, p. 167.

p. 47   "Don't you think caution . . ." Quoted in Moore, p. 167.

p. 49–50   "Bohr's role . . ." Quoted in *Niels Bohr: A Centenary Volume*. Edited by A. P. French and P. J. Kennedy. Cambridge, Mass.: Harvard University Press, 1985, p. 9.

p. 51   "Just the renunciation . . ." Quoted in Moore, p. 158.

p. 51–52   "The main obstacle . . ." Quoted in Moore, p. 218.

4

Splitting the Atom

LISE MEITNER AND NUCLEAR FISSION

Lise Meitner studied beta radiation, then helped discover nuclear fission. She also struggled to survive as the Nazis overran Europe. (American Institute of Physics, Emilio Segrè Visual Archives)

. . . .

Ernest Rutherford had used alpha particles to knock proton "chips" off oxygen nuclei and turned them into nitrogen. Rutherford's work had revealed the potential energy locked up in the atomic nucleus, but it was not clear what other types of collisions might do to the nucleus.

Marie Curie had, through force of personality and exhausting work, made science accept that a woman could be an equal at the

KEY CONCEPTS

binding energy the energy that keeps the particles in an atomic nucleus from flying apart due to electrostatic repulsion

chain reaction the result of neutrons from one atomic fission hitting nearby nuclei and triggering more fissions that release still more neutrons

critical mass the amount and arrangement of material necessary to create a nuclear chain reaction

cyclotron a device that uses electromagnetic fields to accelerate charged particles, or ions, around a circular track

fission the process by which a neutron hits a heavy nucleus (such as uranium), splitting it into two pieces

isotope a variety of an element that has a particular atomic weight

neutron an uncharged particle about the size of a proton. It helps hold the nucleus together.

plutonium an artificial element that is radioactive and can be used for nuclear fission

reactor a device that uses controlled nuclear fission to create power or useful radioactive substances

uranium 235 the scarce isotope of uranium that most easily participates in fission

highest levels of achievement. During the 1930s, another woman, Lise Meitner, would contribute key insights to the understanding of what happens when a nucleus is not merely chipped, but split in two. And while Meitner and other Jewish physicists fled Hitler's domination of Europe, a secret project in the United States would use that knowledge to build a bomb of unbelievable power.

Lise Meitner (1878–1968) was born in Vienna, Austria. The Meitner family, like most European Jewish families, was not unusual in its enthusiasm for education—except for one thing: it applied not only to boys but equally to Lise and the other four daughters. Normally, girls were only educated until age fourteen, at which point they might learn a trade. A bright child like Lise probably found the courses in the girls' school to be boring and unchallenging: some "practical" mathematics; a bit of science; history; geography; French; and such things as drawing, singing, and "feminine handiwork." Later, Meitner would recall that

> Thinking back to . . . the time of my youth, one realizes with some astonishment how many problems then existed in the lives of ordinary young girls, which now seem almost un-imaginable. Among the most difficult of these problems was the possibility of normal intellectual training.

The only intellectual career open to girls was teaching subjects that did not require a university degree. Lise was already interested in physics and mathematics. She had been inspired by reading about Marie Curie's work and was intrigued by the strange world of radioactivity. But Lise's chance of a career in physics must have seemed as remote as becoming a star basketball player today.

Finally, though, it was announced that women would be admitted to the universities if they passed an examination called the Matura. Lise's education from girls' school was woefully inadequate by college standards. Lise, like many other young women, joined with a small group of women who studied together under a tutor. They crammed the equivalent of eight years of high school and college undergraduate work into only two years. Lise passed the test and entered the University of Vienna in 1901, where she

CONNECTIONS

NEW OPPORTUNITIES FOR WOMEN

When Austria gained control of Bosnia-Herzegovina in 1908 officials found they had to provide female doctors for the largely Muslim province because Muslim custom did not allow men to examine women's bodies. As a result, training programs were started for women doctors, and the publicity from them encouraged other women such as Gisela Meitner, Lise's sister, to seek medical training. It was now hard for government and university officials to admit women into medicine and still claim that the female mind was unsuited to other scientific and academic careers.

plunged into courses in calculus and physics. She found the lecture rooms and laboratories to be almost unusable, but the teachers turned out to be excellent.

After receiving her doctorate at Vienna in 1906, she moved to the University of Berlin. She studied under Max Planck (discoverer of the quantum of radiation) and began her own work in radioactivity. She soon met a young chemist named Otto Hahn and wanted to work with him at the university's Chemical Institute. However, Emil Fischer, the head of that institute, did not want any women in the main laboratory. When Max Planck protested on behalf of Meitner, Fischer relented only far enough to allow Meitner and Hahn to set up a small lab in the basement carpentry shop. As biographer Armin Hermann noted

> [Dr. Meitner] was not allowed—heaven only knows why—to enter the student laboratories on the upper floors. Perhaps the privy councillor [Fischer] feared she would distract his students—she was certainly not bad looking. The first women in science could hold their ground in a man's world only by being fastidiously businesslike.

Yet Meitner would have fond recollections of those days:

> When our own work went well we sang together in two-part
> harmony, mostly songs by Brahms. I was only able to hum, but
> Hahn had an outstanding voice. We had a very good scientific
> and personal relationship with our young colleagues at the
> neighboring Physical Institute. They often came to chat and
> would sometimes climb in through the window of the "carpen-
> try shop" instead of taking the usual way. In short, we were
> young, happy, and carefree—perhaps politically too carefree.

As with Marie and Pierre Curie, Meitner's and Hahn's skills fit
well together. Meitner had the training in physics, while Hahn
was a superb chemist. In 1912, Meitner and Hahn moved to the
new and well-equipped Kaiser Wilhelm Institute in Berlin, where
they no longer had to work in a "carpentry room." The year 1914
brought World War I, and Hahn did war research while Meitner
worked as an X-ray technician in military hospitals. Pursuing their
research when they could, they announced in 1918 the discovery
of a new chemical element, protactinium. The name came from
proto-actinium or "before actinium," because the element broke
down into the slightly lighter actinium.

The discovery brought Meitner and Hahn prizes from both the
Berlin and Austrian academies of science. Meitner's achievements
broke down the remaining barriers to an academic career. In 1922,
she finally received "habilitation," the certificate that allows someone to teach in German universities as a full-fledged professor.

Meitner then focused on beta decay: the form of radioactive
breakdown in which an atom sends out a beta particle—that is, an
electron—and gains a proton, moving up one step in the periodic
table. She became involved in a scientific controversy about the
true nature of beta decay. There were two kinds of beta spectra,
one with lines and one continuous. Physicists hoped to use these
spectra to answer the question of whether the electron that was
emitted as a beta particle came somehow from the nucleus of the
radioactive atom, or was an orbiting electron that was jarred loose
by the energy of atomic decay.

Considering the line spectrum, Meitner believed that it resulted from an electron coming out of the nucleus and colliding with orbiting electrons, leaving a kind of "energy fingerprint" that could be interpreted by the quantum theory. On the other side of the debate was a physicist named C. D. Ellis at Cambridge University, England, who thought that the beta electron came from the orbiting electrons, not the nucleus.

Ellis designed an experiment that allowed all the energy in beta decay to be converted to heat that could be measured. The amount of energy recorded seemed to support Ellis, but Meitner remained convinced that some energy was escaping in the experiment—perhaps in the form of gamma rays—and not being measured. While the gamma rays were never found, the methods that Meitner used later formed the basis for Wolfgang Pauli to suggest that the missing energy was in the form of a neutrino, a virtually massless as well as chargeless particle that would be detected in the 1930s.

A Shadow Falls

As the 1930s began, Meitner had become one of the top nuclear physicists in the world, specializing in the analysis of events in the atomic nucleus. Albert Einstein called her "the German Marie Curie." But the days of being "politically carefree" were ending. Hitler's Nazi party came to power in Germany in 1933, and Meitner's license to teach at the University of Berlin was revoked because she was Jewish. At first her ability to do research was not threatened because she was an Austrian citizen.

Despite the political storms brewing around her, Meitner had been able to continue her work with Otto Hahn. She recalled that

> There was a strong feeling of solidarity among us. It was built in mutual trust and this made it possible for the work to continue quite undisturbed even after 1933, although the staff was not entirely united in its political views.

While there was some talk of resistance, many scientists proved no more courageous than many other people in Nazi Germany. When a protest against the treatment of Jews was suggested by Otto Hahn, Max Planck and others advised, "If today you assemble 50 such people [protestors], then tomorrow 150 others will rise up who want the positions of the former, or who in some way wish to ingratiate themselves with the Minister."

In March 1938, Hitler's troops marched into Austria, forcibly attaching the country to Germany. Now Austrians were subject to the full force of the Nazi race laws, and Jews like Meitner were stripped of their rights, their livelihood, and soon, their freedom.

In Denmark, Niels Bohr had recognized the Nazi threat from the very first. The Danes established "The Danish Committee for the Support of Intellectual Workers in Exile." During a trip to Germany, Bohr urged Otto Frisch, a young physicist and Lise Meitner's nephew, to come to Denmark.

Meanwhile, Meitner's friends tried to get her visa papers so she could leave Germany. Her passport, issued by Austria, was no longer recognized by other nations, and the German government would not issue her a new one. Finally, they learned of a border crossing that might let her get into the Netherlands (Holland) without a visa. She spent a last night at Otto Hahn's house. Hahn recalls:

> We agreed on a code-telegram in which we would be let known whether the journey ended in success or failure. The danger was in the SS's [Nazi storm troopers] repeated passport control of trains crossing the frontier. People trying to leave Germany were always being arrested on the train and brought back. . . . We were shaking with fear whether she would get through or not.

Meitner's luck held, and she made it safely across the border. From the Netherlands, Meitner went to Sweden, where she found work at the Nobel Institute of Theoretical Physics at Stockholm. But without her friends and colleagues of many years, Meitner became lonely and depressed. In one letter she wrote: "I feel like a wind-up doll that automatically does certain things, gives a friendly smile, and has no real life in itself." But science would soon catch up with her in the form of a startling discovery.

Discovering the Neutron

In 1920, Rutherford had predicted that a neutral (uncharged) atomic particle would be discovered to account for some of the mass of the nucleus. In 1930, two German researchers discovered that when beryllium was bombarded by alpha particles, a mysterious uncharged radiation was emitted. In 1932, Irène Curie (Marie's daughter) and her husband, Frédéric Joliot, showed that if this radiation struck paraffin, protons were kicked out of hydrogen atoms in the paraffin. No other form of radiation, not even the energetic gamma rays, was known to do this. It would seem that the new radiation would have to consist of particles about the same mass as a proton.

The mass of an electron or alpha particle can be measured by measuring the amount of outside energy needed to balance its charge. But how can one measure the mass of an uncharged particle? In 1932, the British physicist James Chadwick managed to measure the particles by carefully observing their effects on the nuclei of various atoms—something like finding an unknown planet because of wobbles in the orbit of a known one. The new particle was called the neutron, and it earned Chadwick a Nobel Prize in 1935. Neutrons turned out to be slightly more massive than protons.

The neutron was a very useful particle indeed. First of all, it explained why the atomic weight of many elements is at least two times its atomic number. The atomic number is equal to the number of protons, and the remaining mass is found mainly in the neutrons. All atoms of uranium, for example, have 92 protons, but there are three different versions, or isotopes, of the element, with weights of 234, 235, and 238. The difference in weight is accounted for by the different number of neutrons (142, 143, and 146). These differences would turn out to be vitally important.

An Unexpected Explosion

The protons in an atom's nucleus are positively charged. Ordinarily, having the same charge, the protons would repel one another and fly apart. Obviously, there must be some other force that holds the nucleus together. As physicists tried to measure this force, the neutron turned out to be the master key to the nucleus. If the neutrons were arranged in a certain way within the nucleus, they could keep the protons far enough apart to reduce the repulsive force between them. The neutrons could also carry some of the attractive force.

Neutrons also made excellent "bullets" for bombarding nuclei. As physicist Isidor I. Rabi noted

> Since the neutron carries no charge, there is no strong electrical repulsion to prevent its entry into nuclei. In fact, the forces of attraction which hold nuclei together may pull the neutron into the nucleus. When a neutron enters a nucleus, the effects are about as catastrophic as if the moon struck the earth. The nucleus is violently shaken up by the blow, especially if the collision results in the capture of the neutron. A large increase in energy occurs and must be dissipated, and this may happen in a variety of ways, all of them interesting.

As physicists began neutron bombardments, however, the results were very puzzling. Enrico Fermi in Italy thought that he had created strange new elements with atomic numbers higher than uranium (92), the most massive known element. In Paris, France, the Joliot-Curies did their neutron bombardment and came up with what seemed to be, not heavy elements like radium, but medium-weight nuclei. In Berlin, Otto Hahn and his assistant Fritz Strassman had also been bombarding the uranium nucleus with neutrons. Looking at the result, they thought at first they had "chipped" away enough of the nucleus to produce radium. Uranium (number 92) and radium (number 88) are four atomic numbers apart. It was puzzling, though, that a single neutron

OTHER SCIENTISTS

CURIES: THE NEXT GENERATION

Marie Curie's daughter Irène had a very special education indeed, in a school created by her mother that included physicist Paul Langevin as a teacher. During World War I, Irène helped her mother set up mobile X-ray services; she later worked at the famous Radium Institute.

Frédéric Joliot's background was rather different from Irène Curie's. Frédéric's father owned a hardware store, but he had a rather high social position. A physics course taken under Paul Langevin deeply impressed young Joliot and turned him toward a career in science. Langevin eventually got him a job as an assistant at the Radium Institute. There he met Irène, and they married. Rather unusually for the time, they decided to hyphenate their names, becoming the Joliot-Curies.

The Joliot-Curies were in the forefront of nuclear physics in the 1930s. Their bombardment of beryllium led to the discovery of the neutron by James Chadwick in 1932. In 1934, the Joliot-Curies created the first artificial nuclear isotopes by using alpha rays to bombard light elements such as aluminum, making the nuclei unstable.

could change the number of a nucleus by four; alpha and beta rays, for example, changed it only by one.

As Marie Curie had found, radium and barium were very close chemical relatives, which meant they were always found together and were very hard to tell apart. Hahn's team tried to separate the radium that they thought they had created from the surrounding barium. Hahn, the careful chemist, checked the results again and again, but whatever he did, he could not find anything that could be distinguished from barium. Could the product of the nuclear reaction have been radioactive barium, not radium? "There was nothing in the knowledge of nuclear physics at the time to suggest that barium could possibly be produced as a result of the irradiation of uranium with neutrons," Hahn recalled. "Therefore we could only conclude that the products must be isotopes of radium. Still, it was a strange affair."

If barium, with an atomic number of 56 had been produced, it must mean that the uranium nucleus, rather than being chipped, had been split into two pieces. Still troubled about the results, Hahn wrote to his old friend and colleague Lise Meitner in Sweden, and she immediately wrote back:

> Your radium results are very startling. A reaction with slow neutrons that supposedly leads to barium! . . . At the moment the assumption of such a thoroughgoing breakup seems very difficult to me, but in nuclear physics we have experienced so many surprises, that one cannot unconditionally say: it is impossible.

A Walk in the Woods

Meanwhile, Lise Meitner had decided to spend a winter holiday in the Swedish town of Kungälv with her nephew Otto Frisch, who was also a physicist and had been working with Niels Bohr in Copenhagen, Denmark. Meitner showed the letter containing Hahn's barium results to Frisch. As Frisch recalled:

> The suggestion that they might have after all made a mistake was waved aside by Lise Meitner; Hahn was too good a chemist for that, she assured me . . . We walked up and down in the snow, I on skis and she on foot (she said and proved that she could get along just as fast that way), and gradually the idea took shape that this was no chipping or cracking of the nucleus but rather a process to be explained by Bohr's idea that the nucleus is like a liquid drop; such a drop might elongate and divide itself . . .

During the 1930s, Bohr had worked on his idea that the shape of the nucleus was determined by the balance between electrostatic (repulsive) forces and the attraction that protons and neutrons have for one another when close together. Bohr suggested that a nucleus might behave much like a drop of liquid. (Bohr may have been thinking back to the studies of liquid surface tension he had done as a graduate student.)

Meitner wondered whether in some heavy atoms like uranium the two forces might be much more delicately balanced, making the nucleus like a quivering drop of water hanging from a faucet. When a "slow" neutron enters the nucleus, its energy might set the nucleus vibrating and elongating it—like a drop lengthening as it hangs. If the nucleus is elongated enough, Meitner and Frisch agreed, the particles within might be pulled far enough apart so that the electrostatic, repulsive force overcame the nuclear binding force. The nucleus would then split into two roughly equal portions.

Meitner asked a passing biologist what it was called when a cell split in two. "Fission," he replied. So Meitner and Frisch used the term *nuclear fission* to describe the splitting of the atom.

But there was another factor. Physicists knew that the protons and neutrons in the nucleus had a bit less mass when together than they would have as a group of separate particles. The extra mass existed in the form of the "binding energy" that held the nucleus together. When the nucleus is split, this excess energy is released in a flash. How much energy would that be? Meitner used Einstein's famous equation, $E = MC^2$ (energy equals mass times the speed of light squared) to get the answer. Plugging the mass difference of about a fifth the mass of a proton into the equation produced a result of 200 MeV (million electron volts), just the amount of energy that would be enough to force the fragments apart. This energy made a gram of uranium potentially more powerful than thousands of grams of an explosive like dynamite.

Otto Frisch traveled back to Copenhagen and told Niels Bohr about these results. "Oh, what idiots we have all been!" Bohr exclaimed. "Oh, but this is wonderful! This is just as it must be!"

An exciting and awful possibility then arose. In addition to the two big chunks of nucleus and the energy, the fission reaction

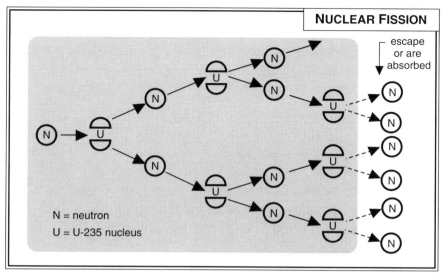

NUCLEAR FISSION

escape or are absorbed

N = neutron
U = U-235 nucleus

In nuclear fission, each splitting nucleus releases two neutrons which in turn can split other nuclei.

created perhaps two to three fresh neutrons. If these in turn could hit and split other uranium nuclei, and each of *these* nuclei spit out two or three neutrons, and so on, the result would be a chain reaction, a toppling of nuclear dominoes that might, if controlled, become a very efficient power source—or if not controlled, a stupendous explosive like nothing the world had ever seen.

What made this last possibility so frightening was that just as scientists in laboratories throughout Europe and the United States were hearing the news about nuclear fission from Meitner and Frisch, Hitler's troops were invading Czechoslovakia. World War II was about a year away.

A Sense of Urgency

The discovery of nuclear fission electrified the scientific world. As laboratories in many countries raced to duplicate Hahn and Meitner's results, Edward Teller, future architect of the hydrogen bomb, felt that there was no time to lose:

> . . . I believe that urgent action [to maintain secrecy] is required. Very many people have discovered already what is involved . . . I repeat there is a chain-reaction mood in Washington. I only had to say "uranium" and then could listen for two hours to their thoughts.

But it was one thing to find a few radioactive barium nuclei that indicated fission had taken place, and another thing to create a chain reaction that could be kept going indefinitely.

Until 1935, scientists thought there was only one naturally occurring isotope of uranium, with a weight of 238, called U-238. Using a mass spectrograph, a device that can detect slight differences in atomic weight, Arthur Dempster of the University of Chicago found tiny traces of another isotope with a weight of 235. U-235 turned out to be present in only about 0.7% of natural uranium.

Niels Bohr took a closer look at how much energy was needed to trigger fission in different nuclei. It turned out that U-235 (and other heavy nuclei with odd mass numbers) released more energy when they absorbed a neutron and changed to an even mass number than did those nuclei that started out even and became odd. This meant that an isotope like U-235 could fission even when hit by a "slow," or lower energy neutron. U-238, on the other hand, could only fission when hit by a fast neutron.

The neutrons that were spit out when fission occurred would be mostly absorbed if they hit U-238 nuclei, but if they hit U-235 nuclei, they would trigger more fissions. Therefore, if one wanted to build the nuclear fire of a chain reaction, the "kindling" would be uranium enriched by having a much higher proportion of U-235 than found in nature. Given a suitable supply of U-235, a chain reaction seemed quite possible.

When Niels Bohr arrived in the United States, he spread news of the fission research of Hahn, Meitner, the Joliot-Curies, and Enrico Fermi to physicists in America. One of them, Leo Szilard, believed that the American government did not yet understand what was at stake. He went to Albert Einstein, who had fled the

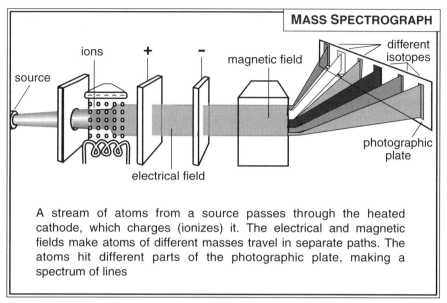

MASS SPECTROGRAPH

ions

+ –

magnetic field

source

different isotopes

photographic plate

electrical field

A stream of atoms from a source passes through the heated cathode, which charges (ionizes) it. The electrical and magnetic fields make atoms of different masses travel in separate paths. The atoms hit different parts of the photographic plate, making a spectrum of lines

A mass spectrograph can be used to separate and measure different isotopes of the same element.

Nazis and taken a position at the Institute for Advanced Study at Princeton University.

Einstein in turn wrote a letter to U.S. president Franklin D. Roosevelt warning:

> It may be possible to set up a nuclear chain reaction by which vast amounts of power and large quantities of radium-like elements might be generated . . . and it is conceivable— though much less certain—that extremely powerful bombs of a new type might be constructed.

The Road to Los Alamos

While the government was slow to react at first, American leaders eventually became convinced that they were in a race with Germany and possibly Russia or Japan to develop nuclear energy and weapons. But just as the United States was far from prepared for

I WAS THERE

A SCARY FLIGHT

When the Nazis occupied Denmark in 1940, they treated the Danes unusually gently at first. This was probably because the light-skinned, blue-eyed Danes were, according to Nazi racial theories, superior "Aryan" types. By 1943, however, the Nazis had begun to treat Denmark more like the other occupied countries of Europe. The Danes, in turn, resisted, particularly when the Nazis demanded that all Jews be turned over to them. The Swedish ambassador warned Niels Bohr (who was Jewish) that he was about to be arrested. The Bohr family was ferried at night by boat across the narrow waterway separating Denmark and Sweden, dodging Nazi patrol boats. Bohr in turn helped organize Swedish efforts to rescue the thousands of Jews who were fleeing across the waters to Sweden.

Bohr was still in danger of being killed or dragged back to Germany by German agents. The British then arranged for Bohr to be flown to Scotland. The ride was far from comfortable. Bohr was crammed into the bomb compartment of a Mosquito bomber that would fly at high altitude to evade German planes. Bohr was given an oxygen mask and a helmet with a radio headphone, but the helmet did not properly fit his large head. He did not hear the instruction to turn on the oxygen and passed out. Fortunately, the plane crew, realizing Bohr was not responding, flew to a lower altitude where he revived. Bohr was soon on his way to the United States to work on the atom bomb project.

the coming world war, nuclear physics in America was still rather primitive by European standards.

One area where Americans were starting to make a contribution, however, was in the building of machines popularly called "atom-smashers" to accelerate particles and slam them into nuclei. One important center for nuclear physics in the United States was the Radiation Laboratory at the University of California, Berkeley, founded by Ernest O. Lawrence, who had invented a machine called the cyclotron.

The cyclotron used electromagnetic fields to accelerate ions (charged atoms) around in a circle. Strong magnets kept the ions from flying off in a straight line, and as they came around again each time a fresh pulse of energy would speed them up some more, until they reached a maximum energy where they could no longer be held by the magnets, and shot off in the form of a beam that could be directed at a target nucleus.

The cyclotron and other types of particle accelerators made it possible to have high-energy particles on demand without having to use expensive radioactive substances or having to wait for a natural high-energy cosmic ray to come zipping in from space. The accelerators marked the beginning of the high-tech, massively funded "big physics" that would be evident by the 1950s.

The 60-inch cyclotron at the Lawrence Berkeley Laboratory (Lawrence Berkeley Laboratory)

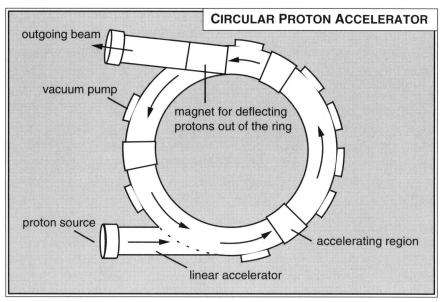

CIRCULAR PROTON ACCELERATOR

outgoing beam

vacuum pump

magnet for deflecting
protons out of the ring

proton source

accelerating region

linear accelerator

A typical cyclotron-type proton accelerator. A linear accelerator is used to inject the protons into the ring.

The nuclear effort that would eventually be given the code name "Manhattan Project" was perhaps the largest, most complex single project in all of human history. According to Herbert S. Marks, "The Manhattan District bore no relation to the industrial or social life of our country; it was a separate state, with its own airplanes and its own factories and its thousands of secrets." Put another way, in only three years, the United States government spent two billion dollars to create from scratch an industry as large as the entire American automotive industry.

The project would have several phases. At Oak Ridge, Tennessee, U-235 was separated from the less energetic U-238. Two methods of separation were used. The first, the "calutron" (named for the University of California) used the cyclotron principle. When electromagnetically accelerated around the "racetrack," the beams of U-235 and U-238 ions would take slightly different paths because of their differing weights. The second method, called gaseous diffusion, relied on the fact that a lighter gas can be pushed farther through a porous membrane (something like a cigarette filter) than a heavier gas. The gas in this case was uranium hexaflouride, and after pumping the gas through a long series of such filters, a high concentration of U-235 hexaflouride could be obtained. The problem was that uranium hexafluoride was a very dangerous, corrosive gas.

The question of uranium separation set the pattern for the whole atom bomb effort. Since scientists and engineers often couldn't be sure which method for doing something would work, the government leaders, increasingly worried about a possible Nazi atom bomb, funded an all-out effort using each possible method. Time, not money, was what worried them most.

I WAS THERE

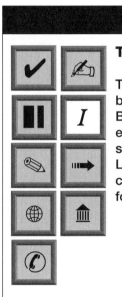

THE COATS OFF THEIR BACKS

Today's high-energy physics laboratories have budgets in the hundreds of millions of dollars. But things were quite different in the Depression-era 1930s. Philip H. Abelson, while a graduate student and research assistant to Ernest O. Lawrence at Berkeley, had to make a difficult choice. He needed about 5 kilograms of uranium for an experiment. But, as he recalled:

> In those days the laboratory had little money, especially for graduate students. The major funds went into running and improving the cyclotron. I desperately wanted the uranium. How to get it? My stipend [expense allowance] was $60 a month. One day I received a letter from my parents enclosing money to buy a new suit. My wife and I went to San Francisco to get it; but before we arrived at the store I saw the sign of Braun Knecht and Heiman, vendors of chemicals. The money for the suit was diverted to the purchase of uranium oxide.

Meanwhile, a separate laboratory at the University of Chicago went to work building the world's first nuclear reactor. There were two important reasons for doing so. First, scientists needed to learn how to control the fission reaction. There are three possible conditions when dealing with fissionable material. In the first, a few fissions may be occurring here and there, but most of the neutrons released by the splitting atoms are escaping or being

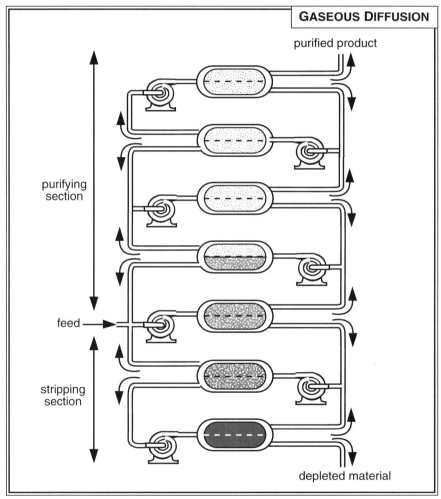

A gas diffusion apparatus was used to gradually separate the lighter U-235 gas from the heavier U-238.

absorbed by surrounding material that can't itself be fissioned. There is therefore no increase in the rate of fission.

In the second condition there is a balance: just enough neutrons are being produced and creating new fissions so that fissions are occurring at a steady rate, as in a nuclear power plant. The amount of fissionable material (U-235, for example) necessary to do this is called "critical mass."

Finally, if enough neutrons are starting new fissions so that one fission produces two or more other fissions, which produce double *that* number of fissions, and so on, very quickly—the result is an uncontrolled nuclear reaction and the blast of an atomic bomb.

The scientists in Chicago, including Lise Meitner's nephew Otto Frisch, needed to learn all the factors that went into creating and sustaining a chain reaction. These included the shape and spacing of the fissionable material, the moderator (a material such as water or graphite used to slow down neutrons so they are more likely to cause new fissions), and the position of rods of neutron-absorbing material that act as a kind of brake on the reaction process.

The world's first nuclear reactor was built under the stands of the University of Chicago football stadium. It used hundreds of bricklike blocks of graphite, in which uranium was embedded. On December 2, 1942, the reactor went critical—that is, produced a self-sustaining chain reaction—for the first time.

But there was another use for reactors. The analysis of neutron energy suggested that in addition to U-235, plutonium (Pu-239), an artificial element recently created in the laboratory, could also be used to make a chain reaction or a nuclear bomb. Researchers discovered that U-238, otherwise useless for fission, could be bombarded by neutrons and become U-239, which eventually decays to Pu-239, which is fissionable. A set of five nuclear reactors were built in Hanford, Washington, and used to produce Pu-239 for nuclear weapons.

There were basically two different designs for the bombs themselves. The first used U-235 and worked something like a cannon stuck into a tube. The cannon had as its "shell" a piece of half the critical mass of highly enriched U-235; a second piece is seated in

I WAS THERE

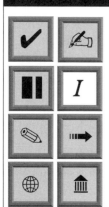

A CLOSE CALL

Otto Frisch, codiscoverer of nuclear fission (with Lise Meitner), played a major role in the nuclear bomb research at Los Alamos during World War II. One of his experiments consisted of a spherical apparatus used to test calculations of the critical mass—the amount and arrangement of fissionable material that results in a self-sustaining chain reaction. Frisch recalls:

> On one occasion I was making such an assembly . . . and just as we were getting close to critical size the student who was assisting me pulled out the neutron counter which he said seemed unreliable. I leaned over, calling out to him to put it back, and from the corner of my eye I saw the neon lamps on the scaler [radiation meter] had stopped flickering and seemed to glow continually. Hastily I removed a few pieces of uranium-235, and the lamps returned to their flickering. Obviously, the assembly had briefly become critical because my body—as I leaned over— reflected neutrons back into it. By measuring the radioactivity of some of the uranium-235 bricks afterwards, we could calculate that the reaction had been growing by a factor of 100 every second! As it happened, I had received only about one standard daily dose [of radiation] in those two seconds, but it would have been a lethal dose if I had hesitated for two seconds longer.

Others were not so lucky. Two later accidents with the uranium-235 bricks resulted in the death of researchers.

the end of the tube, and the pieces are shaped so they will fit together. The firing of the gun mechanism brings the two pieces together in an instant, creating a critical mass, and the pressure on the material concentrates the neutrons and causes an uncontrolled chain reaction—a nuclear explosion.

The second type used Pu-239 and a more sophisticated mechanism. A spherical shell of plutonium was surrounded by explosives

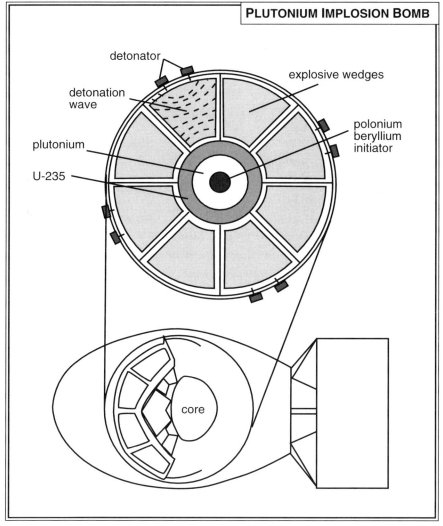

PLUTONIUM IMPLOSION BOMB

detonator

detonation wave

explosive wedges

plutonium

polonium beryllium initiator

U-235

core

The implosion-type plutonium bomb directs explosive waves inward toward the core.

shaped so that when they were set off the plutonium would be "imploded" inward into a super-dense critical mass that would then cause a nuclear explosion.

The bombs were built and tested in a top secret facility near Los Alamos, New Mexico. The first test, using the plutonium bomb and held at a site called Trinity, took place on July 16, 1945. And on August 6, a U-235 atomic bomb destroyed the Japanese city of Hiroshima. Three days later, a PU-239 bomb was dropped on another Japanese city, Nagasaki. Together, the bombs killed about 150,000 people, wounded about an equal number, and condemned thousands to die years later from radiation-related illnesses.

In the years to come many people would question the need to use such a devastating weapon on a nation that was already close to defeat. On the other hand, American leaders argued at the time that forcing Japan to surrender would save hundreds of thousands of American and Japanese lives that otherwise might have been lost in a bloody invasion of Japan by American forces.

Meanwhile, Lise Meitner, whose work had helped make all of this possible, had quietly faded into the background. In 1944, Otto Hahn was given the Nobel Prize in chemistry for the discovery of nuclear fission. While German scientists celebrated, Meitner's friends in Sweden were dismayed that she had not been given a share of the prize. Birgit Broome Aminoff, a scientist and wife of a member of the Nobel Prize committee wrote to Meitner:

> Long before the release of nuclear energy had been realized on a practical scale, it seemed to me that Professor Meitner had reached a status equivalent to that of many Nobel Prize recipients. It must therefore have been very bitter that for completely unrelated reasons you were forced to leave the laboratory where the now-rewarded discovery was so close, and thereby lost the possibility to complete a work which promised to be the natural climax of a long and devoted career as a scientist.

Be that as it may, Lise Meitner had received many honors by the time she died in 1968, including the prestigious Enrico Fermi

Award. Starting in 1940 with neptunium and plutonium, researchers had learned how to make artificial elements with higher atomic numbers than uranium. In 1982, German scientists created one such element with atomic number 107 and named it meitnerium in honor of Lise Meitner.

Meitner's life was marked by hard work and quiet integrity. Looking back, she addressed a new generation by saying:

> I believe all young people think about how they would like their lives to develop. When I did so, I always arrived at the conclusion that life need not be easy provided only that it was not empty. And this wish I have been granted.

In the midst of social turmoil, rapid scientific progress, and atomic dangers, Meitner had grounded herself in the basic things of life. The inscription on the headstone of her grave reads simply:

> Lise Meitner; a physicist who never lost her humanity.

Chronology of Lise Meitner and Nuclear Fission

1907	Lise Meitner begins work with Otto Hahn in Berlin
1918	Meitner and Hahn discover the element protactinium
1922	Meitner becomes a full-fledged professor
1932	James Chadwick discovers the neutron Werner Heisenberg shows how neutrons in the nucleus explain the existence of isotopes of the same element
1934	Irène Joliot-Curie and Frédéric Joliot-Curie create first artificially radioactive isotope
1936	Bohr develops "liquid drop" model of the atomic nucleus

1938   Germans take control of Austria; Meitner flees to Sweden

1939   Lise Meitner and Otto Frisch publish paper on nuclear fission
        Albert Einstein warns President Franklin Roosevelt about the possibility of nuclear weapons

1942   Manhattan Project begins; first chain-reaction reactor tested in Chicago

1945   Nuclear fission weapon (atomic bomb) tested and dropped on Hiroshima and Nagasaki

1966   Lise Meitner wins the Enrico Fermi Award

1982   Scientists name a new element, meitnerium, in honor of Lise Meitner

Further Reading

All In Our Time: *The Reminiscences of Twelve Nuclear Pioneers.* Edited by Jane Wilson. Chicago: Bulletin of the Atomic Scientists, 1974. Vivid recollections by people involved in nuclear research in the 1930s and the World War II atomic bomb project.

Heilbron, J. L. and Seidel, Robert W. *Lawrence and His Laboratory: A History of the Lawrence Berkeley Laboratory.* Berkeley: University of California Press, 1989. The story of scientists and their discoveries at one of the key institutions in nuclear research.

Rhodes, Richard. *The Making of the Atomic Bomb.* New York: Simon and Schuster, 1986. A detailed but readable history of the massive American atomic bomb effort during World War II.

Yount, Lisa. *Twentieth-Century Women Scientists.* New York: Facts On File, 1996. Contains a chapter on Lise Meitner; written for young adults.

NOTES

p. 57   "Thinking back to . . ." Quoted in Sime, *Lise Meitner: A Life in Physics*. Berkeley, Calif.: University of California Press, 1997, p. 7.

p. 58   "[Dr. Meitner] was not allowed . . ." Quoted in *The New Physics*, p. 41.

p. 59   "When our own work . . ." Quoted in *The New Physics*, p. 41.

p. 60   "There was a strong feeling . . ." Quoted in Moore, p. 222.

p. 61   "If today you assemble . . ." Quoted in Sime, pp. 145–46.

p. 61   "We agreed on a code-telegram . . ." Quoted in Sime, p. 204.

pp. 61–62 "I feel like a wind-up . . ." Quoted in *The New Physics*, p. 93.

p. 63   "Since the neutron . . ." Quoted in Rhodes, Richard. *The Making of the Atomic Bomb*. New York: Simon and Schuster, 1986, p. 209.

pp. 64–65 "There was nothing . . ." Quoted in Moore, p. 223.

p. 65   "Your radium results . . ." Quoted in Sime, p. 235.

p. 65   "The suggestion that they . . ." Quoted in Sime, p. 236.

p. 66   "Oh, what idiots . . ." Quoted in Frisch, Otto. *What Little I Remember*. Cambridge: Cambridge University Press, 1979, p. 116.

p. 68   "I believe that urgent . . ." Quoted in Rhodes, p. 290.

p. 69   "It may be possible . . ." Quoted in Adrian Berry. *The Book of Scientific Anecdotes*. New York: Prometheus Books, 1993, pp. 173–74.

p. 72   "The Manhattan District . . ." Quoted in Rhodes, p. 277.

p. 73   "In those days . . ." Quoted in *All In Our Time: The Reminiscences of Twelve Nuclear Pioneers*. Edited by Jane Wilson. Chicago: Bulletin of the Atomic Scientists, 1974, p. 27.

p. 76   "On one occasion . . ." Quoted in *All In Our Time*, pp. 61–62.

p. 78   "Long before the release . . ." Quoted in Sime, p. 326–27.

p. 79   "I believe all young people . . ." Quoted in Rhodes, p. 233.

p. 79   "Lise Meitner . . ." Quoted in Sime, p. 380.

5

Picking the Nuclear Lock

RICHARD FEYNMAN'S QUEST FOR UNDERSTANDING

Richard Feynman had boundless curiosity and the ability to translate mathematics into visual terms. (Photograph by Floyd Clark, courtesy of Archives, California Institute of Technology)

Ａs the 1940s began, nuclear physics had changed from a scientific curiosity to a vitally important source of technology for power generation, propulsion, and weapons. As the science changed, so did its geography. Most of Europe's top physicists had fled Hitler's growing empire; many would soon come—or be invited to—the United States. America would replace Europe as the center of nuclear physics research.

One young American physics graduate would perhaps best represent the distinctive personality of the new American physicist. Like many of his European colleagues, he was Jewish. But the accent, the energy, and the attitudes of Richard Feynman would come not from the coffee houses of Vienna or the great institutes of Berlin but from the streets of New York.

Feynman would play a role in the birth of the atom bomb and the nuclear age. He would go on to develop key tools for understanding the world of forces within the atomic nucleus—and the strange new particles that were emerging from ever more powerful atom-smashers.

Learning How to See

Richard Feynman (1918–88) was born in Far Rockaway, then a suburb of New York City, now a section of Brooklyn. His father, Melville Feynman, had already declared that "If it's a boy he'll be a scientist." Certainly Richard's father was his first and perhaps best teacher. When "Ritty" was still very small, Melville obtained a collection of colored bathroom tiles from a company's surplus stock. He arranged them in long rows like dominoes and let his son knock over one end of the arrangement. The boy delighted in this operation, which perhaps foreshadowed the chain reactions that he would be concerned with at Los Alamos. Melville also created patterns of colors with the tiles and challenged his son to continue them.

Richard's father also taught the essence of science in another way. One of Richard's young friends asked him the name of a particular bird. When Richard said he didn't know, the other boy said, "It's a brown-throated thrush. Your father doesn't teach you anything!"

But Richard knew this wasn't true. His father had already told him about the difference between naming a thing and truly understanding it:

> You can know the name of that bird in all the languages of the world, but when you're finished, you'll know absolutely nothing whatever about the bird. You'll only know about humans in different places, and what they call the bird. So let's look at the bird and see what it's *doing*—that's what counts.

In his thinking about physical phenomenon, Richard Feynman would never forget to look at "what the bird is *doing*."

After high school, Feynman was accepted at the Massachusetts Institute of Technology (MIT), a school famous as the creator of

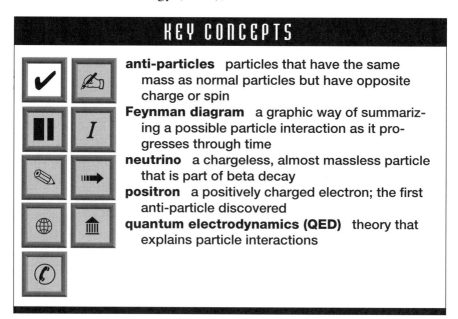

KEY CONCEPTS

anti-particles particles that have the same mass as normal particles but have opposite charge or spin

Feynman diagram a graphic way of summarizing a possible particle interaction as it progresses through time

neutrino a chargeless, almost massless particle that is part of beta decay

positron a positively charged electron; the first anti-particle discovered

quantum electrodynamics (QED) theory that explains particle interactions

America's finest engineers. At first, however, Feynman wanted to be a mathematics major. He gradually became dissatisfied with his mathematics course because the math was abstract and didn't seem to be useful for anything in the real world. He went to the head of the mathematics department and asked him: "Sir, what is the use of mathematics if not to teach more mathematics?" He replied that Feynman could always get a job with an insurance company calculating insurance rates. But Feynman realized that what he really wanted to do was explore the secrets of the real, physical world of nature.

In the physics laboratory Feynman was given a simple-looking experiment. A metal ring was hung from a nail on the wall. The experiment was to "measure the period [time it took to swing back and forth], calculate the period from the shape, and see if they agree." He took an unexpected pleasure from this experiment:

> I thought that was the best doggone thing. I liked the other experiments, but they involved sparks and other hocus-pocus, which was too easy. With all that equipment you could measure the acceleration due to gravity. The remarkable thing is that physics is so good, in that not only can you figure out something carefully prepared but something so natural as a lousy old ring hanging off a hook—that impressed me!

Feynman began to read widely about the latest discoveries in physics. He soon noticed that many important papers in physics were coming from Princeton University, but there were few from MIT. MIT was a first-rate engineering school, but somewhat behind in theoretical physics. So Feynman decided to go to Princeton for his graduate studies. There he became a research assistant to John Wheeler, who would become an important physicist. They quickly found that they worked well together. Because Wheeler was involved in research in new developments in nuclear physics, Feynman focused his attention on the atomic world. He was soon up to speed on quantum mechanics.

The quantum mechanics of the 1920s had gone a long way toward being able to explain the orbital paths of single electrons.

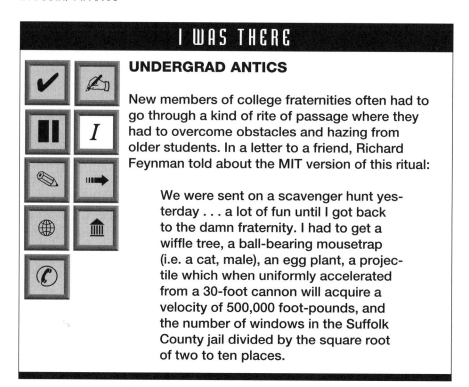

UNDERGRAD ANTICS

New members of college fraternities often had to go through a kind of rite of passage where they had to overcome obstacles and hazing from older students. In a letter to a friend, Richard Feynman told about the MIT version of this ritual:

> We were sent on a scavenger hunt yesterday . . . a lot of fun until I got back to the damn fraternity. I had to get a wiffle tree, a ball-bearing mousetrap (i.e. a cat, male), an egg plant, a projectile which when uniformly accelerated from a 30-foot cannon will acquire a velocity of 500,000 foot-pounds, and the number of windows in the Suffolk County jail divided by the square root of two to ten places.

But Feynman discovered that physicists had run into a mathematical muddle when they tried to explain how electrons interacted in electromagnetic fields. Infinite quantities kept cropping up in their equations. For example, if an electron is a charged round particle with radius a and charge e, its energy is approximately e^2/a. But Bohr's idea of complementarity had shown that the wave and particle view of the electron were intertwined. What happens when one's equation goes from an electromagnetic wave to a particle that is a geometric point with radius 0? One ends up dividing by zero, which gives an infinite (or, strictly speaking, undefined) result. Other more complicated infinities also turned up when electrons were considered to be "harmonic oscillators"—points that vibrated to create the electromagnetic field. One ended up with an infinite number of points.

What was worse, Heisenberg's uncertainty principle allowed "virtual particles" to exist for periods of time too short to measure.

These particles, while undetectable, had to be taken into account in measuring things like the mass of an electron. The theory also made it seem necessary that an electron interact with *itself*.

Feynman attempted to get around this difficulty by saying that an electron could not act on itself but only on other particles. That dispensed with the field with infinite vibrations and potentially infinite energy; there were just particles that directly acted on one another. If one looks toward the sun, Feynman suggested, "The sun atoms shakes; my eye electron shakes eight minutes later, because of a direction across [space]." But this "action at a distance" seemed to be a throwback to pre-19th-century physics, introducing as many problems as it solved. Still, Feynman had entered the heady world where physicists were trying to explain not only particles but the very forces and transformations that gave rise to them.

The Safe-Cracker

Suddenly, Feynman, like other physics graduates, found himself abruptly brought down to earth. By the summer of 1941, it was clear that war was coming. What now? Would he simply be drafted like millions of other young men? Or would physics actually become part of the war effort?

Feynman had been brought into the "real world" in another way as well. He had overcome his shyness about girls and had fallen deeply in love with Arline Greenbaum, an attractive young woman with considerable talent in music and art. Her direct approach to life impressed him, even though her cultured background often clashed with his rationalism and disdain for subjects that he considered to be undisciplined and muddled. But a shadow hung over their relationship: Arline suffered from what turned out to be a form of tuberculosis, a bacterial infection that led to recurring fevers and a slow wasting away.

Feynman went to work for the summer in an army laboratory on a mechanical gun-aiming computer. Meanwhile, the news about nuclear fission and its possibilities was percolating through the government and military establishment. When Feynman returned to Princeton to finish his doctoral thesis, Robert R. Wilson took him into his office and told him that he was forming a team of physicists to build an atomic bomb. Feynman, preoccupied by his thesis and disillusioned with war work, turned him down at first. He then thought about whether Werner Heisenberg and the other physicists remaining in Germany were working on such a bomb. That possibility changed Feynman's mind. Feynman signed up for what would become the Manhattan Project (see Splitting the Atom).

In 1943, Feynman went to Los Alamos, the secret city in the New Mexico desert, where the atom bombs would be designed, assembled, and tested. He found that he was more in demand as a mathematician than as a physicist. He had a knack for solving complicated calculus problems when people with just as much formal training didn't know where to begin. The atom bomb project, with its need to predict the behavior of neutrons under rapidly changing conditions, had a tremendous appetite for numbers.

Feynman soon became involved with organizing and speeding up the process, by which teams of clerks cranked through problems on primitive calculators that could have been done in a few minutes by today's desktop computers. When the mechanical calculators that looked like overgrown typewriters filled with keys and gears kept breaking down, Feynman took one apart and learned how to fix them. There was no time to send them back to the factory to be repaired.

Feynman also learned the most efficient ways to break down complex problems into a series of steps that could be performed by the people—mostly scientists' wives—who worked the calculators. In essence, he created a language and recipes, or algorithms, for solving problems on a "computer" whose chips were human beings.

Feynman's love of finding solutions to all sorts of puzzles showed itself in other ways. He wrote to Arline and admitted that he had probably become obsessed with the many locks that guarded the secret papers around the laboratories:

because I like puzzles so much. Each lock is just like a puzzle you have to open without forcing it. But combination locks have me buffaloed. You do too, sometimes, but eventually I figure out you.

Soon he mastered the combinations as well. When people forgot the combinations to the safes containing their portion of atomic secrets, they came to Feynman. Feynman seemed to be able to get into any safe. Sometimes he could guess the combination or get it by trying some likely possibilities. (When he learned that the safe could not distinguish between the "correct" number and one up to two places away, he could cut down on the number of combinations to try.) He also learned how to turn the knob until the bolt dropped, yielding the combination's last number.

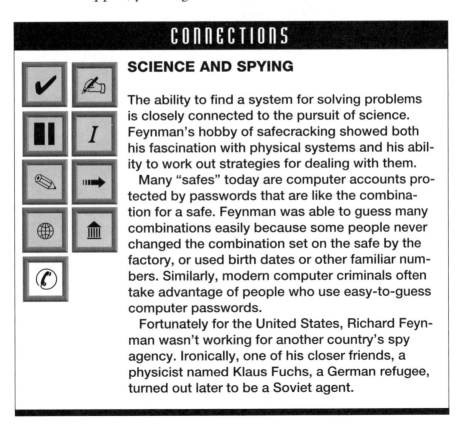

CONNECTIONS

SCIENCE AND SPYING

The ability to find a system for solving problems is closely connected to the pursuit of science. Feynman's hobby of safecracking showed both his fascination with physical systems and his ability to work out strategies for dealing with them.

Many "safes" today are computer accounts protected by passwords that are like the combination for a safe. Feynman was able to guess many combinations easily because some people never changed the combination set on the safe by the factory, or used birth dates or other familiar numbers. Similarly, modern computer criminals often take advantage of people who use easy-to-guess computer passwords.

Fortunately for the United States, Richard Feynman wasn't working for another country's spy agency. Ironically, one of his closer friends, a physicist named Klaus Fuchs, a German refugee, turned out later to be a Soviet agent.

The world's first nuclear explosion affected its observers in a variety of ways. It was clear, however, that a new era had begun. (Los Alamos National Laboratory)

Finally, the atomic bomb was ready to be tested. Lying down and looking through dark glass, the Los Alamos scientists waited in the predawn light while the bomb sat in a steel tower 20 miles away. "And then, without a sound, the sun was shining; or so it looked," Otto Frisch would recall later. Another physicist, Isidor Rabi, wrote of the light: "It blasted; it pounced; it bored its way into you. It was a vision which was seen with more than the eye." Then came a crack like a rifle shot, the rumble of thunder in the air, and finally the wind pushed ahead of the shock wave.

As Feynman wrote to his mother, "We jumped up and down, we screamed, we ran around slapping each other on the backs." It had worked! But the blazing flash of the bomb, "brighter than a thousand suns," made Robert Oppenheimer, director of the laboratory, think of a description of the terrible goddess Kali in the Hindu scriptures of India: "Now I am become death, the destroyer of worlds."

A Return to Theory

Feynman had also met death personally. In the last months before the bomb test, Arline died of a disease that in a few more years would have been curable by antibiotics. Feynman's time at Los Alamos had been one of excitement and shared purpose. The war had been won, but what kind of future might there be in the new atomic world? He was pessimistic about that:

> Most was known . . . Other people are not being hindered in the development of the bomb by any secrets we are keeping. They might be helped a little by mentioning which of two processes is found to be more efficient, & by our telling them what size parts to plan for—but soon they will be able to do to Columbus, Ohio, and *hundreds* of cities like it what we did to Hiroshima.
>
> And we scientists are clever—too clever—are you not satisfied? Is four square miles [destroyed] in one bomb not enough? Men are still thinking. Just tell us how big you want it!

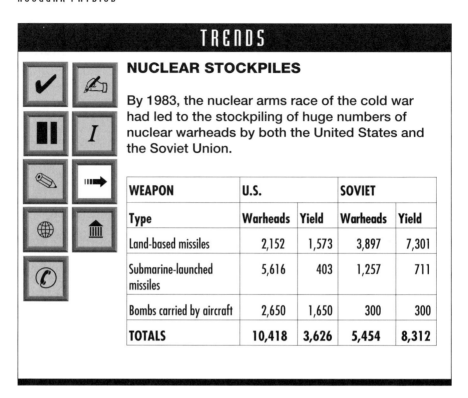

TRENDS

NUCLEAR STOCKPILES

By 1983, the nuclear arms race of the cold war had led to the stockpiling of huge numbers of nuclear warheads by both the United States and the Soviet Union.

WEAPON	U.S.		SOVIET	
Type	Warheads	Yield	Warheads	Yield
Land-based missiles	2,152	1,573	3,897	7,301
Submarine-launched missiles	5,616	403	1,257	711
Bombs carried by aircraft	2,650	1,650	300	300
TOTALS	10,418	3,626	5,454	8,312

Edward Teller would soon be seeking people to work on the hydrogen fusion bomb, the one that would bring hydrogen atoms together and release a hundred times the energy of the Hiroshima bomb. Feynman, however, accepted a job at Cornell University, teaching and trying to find his way back to theoretical physics.

Physics itself seemed to be catching its breath and struggling to find a comprehensive theory in which to fit the discoveries of the 1930s and 1940s. "The theory of elementary particles has reached an impasse," Victor Weisskopf wrote.

During the 1930s researchers had begun to take advantage of nature's own particle accelerator. Atomic nuclei zip into the earth's atmosphere, the product of the sun and perhaps of far-off stellar explosions called novas and supernovas. These "cosmic rays" hit nuclei of atoms in the atmosphere and produce showers of particles that can be seen in a cloud chamber. This device expands gas (such as air with water vapor) by pulling out a piston. As the gas expands, it cools and a cloud forms around ions in the gas. Any particles coming through the chamber leave tracks in the cloud, something like the contrail of a jet plane.

Analyzing such tracks led to some surprises. One of them, predicted by theory, was the positron, a particle that has the mass of an electron but with a positive rather than a negative charge. When a photon (light quantum) with the right amount of energy hits a nucleus, its excess energy creates a pair of particles—an electron and a positron. This illustrated the way that mass and energy could be interchanged. Gradually, it was discovered that

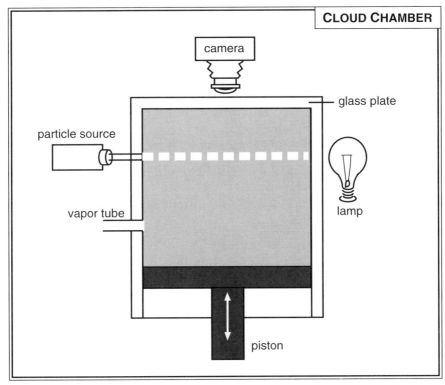

A cloud chamber reveals the tracks of particles within gas vapor.

each elementary particle had its opposite twin, an anti-particle that had an opposite charge—or in the case of the neutron, an opposite spin.

In peering into the world within the nucleus where particles and energy turned into one another, physicists were now investigating the actual forces within the atom. One, called the weak nuclear force, was what made radioactive atoms break down and send out beta particles. The question that Meitner had debated—whether beta particles come from the nucleus or the electrons surrounding the atom—was answered. A neutron in the nucleus became a proton plus an electron (beta particle). But the mass of a proton plus an electron doesn't quite add up to that of a neutron. Fermi suggested that this missing mass was emitted in the form of a particle called a neutrino (Italian for "little neutral one").

Because the neutrino had no charge and virtually no mass, it was very difficult to detect. By the 1950s, however, nuclear reactors were in operation—and these were places where a lot of radioactive decay was going on all the time. Researchers set up a tank full of a solution of cadmium chloride and water. After shielding the tank with old battleship armor to keep out any other kinds of particles, they waited for a few neutrinos to hit hydrogen atoms in the water, ricochet, and eventually get captured by a cadmium nucleus, which in turn emits a gamma ray with a particular energy. The neutrino proved to be a very elusive beast: of the 10^{13} neutrons per square centimeter per second poured out by the nuclear reactor, only three per hour collided with atoms in the tank. But that was enough to confirm that they were there.

Neutrinos were the missing ingredient in the weak force of radioactive decay. In 1935, Japanese physicist Hideki Yukawa studied the strong force—the one that holds protons and neutrons together in the nucleus. He predicted that just as light has its energy carrier (the photon), the strong force would also manifest itself as a particle under certain conditions. Experimenters studying cosmic rays found such particles, which became known as mesons because they have a mass between that of the electron and that of the proton. When a fast-moving proton from a cosmic ray or particle accelerator hits a proton at rest, the result can be two protons and a type of meson known as neutral pi meson (or pion for short), a proton and a neutron plus a positive pion, or if the energy is high enough, even two protons plus a whole flock of pions.

With particles, anti-particles, photons, and an assortment of mesons, the possibilities for particle interactions were rapidly

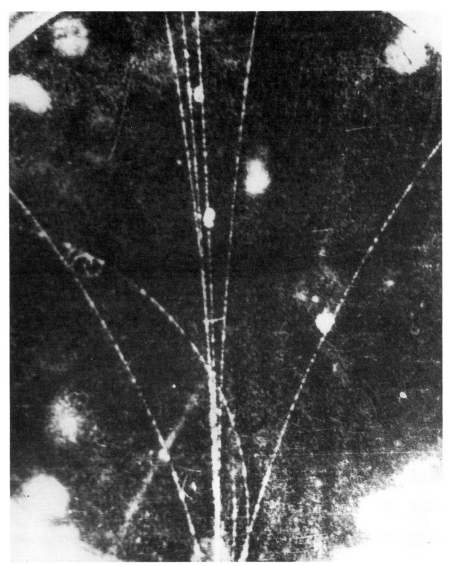

Particle tracks in a cloud chamber (Archives, California Institute of Technology)

growing. Physicists needed a comprehensive way to describe what was going on.

The Atomic Playbook

In Feynman's way of working, mathematics was always tied to a mind picture, a visualization of physical process. As he told an interviewer:

> What I am really trying to do is bring birth to clarity, which is really a half-assedly thought-out pictorial semi-vision thing. I would see the jiggle-jiggle-jiggle or the wiggle of the path.

Feynman found a way to rework the conceptions of quantum mechanics. The standard interpretation of quantum theory, called the "Copenhagen Interpretation" after the physicists in Bohr's group, speaks of many possible paths that particles can take in an interaction. Feynman found a way to mathematically express and sum up these possible paths, or "histories," to get the complete picture of a situation. To aid in visualizing each possible path, Feynman used a diagram to show a possibility in terms of particles moving across the page (representing the passage of time) and exchanging particles such as photons or emitting new particles.

   In 1948, a conference on "fundamental problems of theoretical physics" was held at Pocono Manor, Pennsylvania. Julian Schwinger, an American physicist of Feynman's generation, had worked out a comprehensive and thorough mathematical system for calculating the paths of particles while taking relativity into account. After Schwinger had finished his lengthy presentation, Feynman gave a rather hastily organized talk, in which he introduced his diagrams and rules for using them. Among other things, he suggested that the positron could be treated as though it were an electron moving backward in time. This radical notion probably contributed to the rather poor reception Feynman received. Bohr, in particular, thought Feynman was on the wrong track.

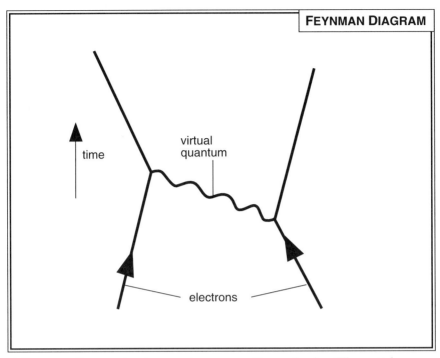

FEYNMAN DIAGRAM

A Feynman diagram showing two electrons repelling each other. The particles move upward as time passes. The virtual quantum carries the energy of repulsion.

By 1949, Feynman was able to give a much more complete presentation in the *Physical Review*. He remarked to Jagdish Mehra:

> In private I had great amusement in thinking that my silly-looking diagrams, when published in the *Physical Review*, would poke fun at that august [dignified] journal. I liked to think that my diagrams were the equivalent of sheep's livers and entrails into which the ancient Greek and Egyptian priests used to look for predicting the future.

Although believers in an elegant, purely mathematical approach scorned Feynman's diagrams at first, students quickly found that they were a valuable aid. It was like having two ways to describe a football play. One way could use sets of numbers showing the velocity and momentum of the quarterback and wide receiver at each instant, plus that of the ball being passed from one to the other. The other way—Feynman's way—related the

I WAS THERE

NATURE HAS THE LAST WORD

In 1986, the space shuttle *Challenger* suddenly blew up shortly after it lifted off the launch pad. Feynman was one of the people appointed to a presidential commission to investigate the accident. The problem was traced to an o-ring, a flexible rubber seal that had to expand quickly enough during flight to seal the rocket joints and prevent a leak of explosive fuel. The various NASA officials, engineers, and contractors involved seemed to be mainly concerned with who had tried to warn of a possible disaster and whether the proper decision-making procedures were followed.

Feynman took a typically direct approach. At one moment in the proceedings he announced:

> I took this rubber from the model and put it in a clamp in ice water for awhile. I discovered that when you undo the clamp, the rubber doesn't spring back. In other words, for more than a few seconds, there is no resilience for this particular material when it is at a temperature of 32 degrees. I believe that has some significance for our problem.

Indeed it had: the weather had been too cold and the o-ring could not expand fast enough to make a good seal. As physicist Freeman Dyson remarked,

> The public saw with their own eyes how science is done, how a great scientist thinks with his hands, how nature gives a clear answer when a scientist asks her a clear question.

mathematics to a diagram like that used by a football coach, with the Xs, Os, and arrows showing what each player is to do.

In 1965, Feynman, Julian Schwinger, and Shin'ichiro Tomonaga shared the Nobel Prize in physics for their contributions to QED, or quantum electrodynamics—the theory that explains particle interactions. Feynman's many-faceted career as teacher, researcher, and self-defined "curious character" would continue until his death in 1988.

Students who have taken or read Feynman's introductory lectures in physics have had a glimpse of the mind of a formidably complex yet disarmingly simple man who was always asking new questions.

Chronology of Richard Feynman and Postwar Nuclear Physics

1936	Existence of force-carrying mesons predicted
1943	Richard Feynman goes to Los Alamos to work on A-bomb project
1948	Feynman introduces space-time diagrams for particle interactions
1956	Neutrinos detected coming from nuclear reactor
1965	Feynman shares Nobel Prize in physics for developments in quantum electrodynamics
1986	Feynman points out o-ring problem following space shuttle *Challenger* disaster

Further Reading

Feynman, Richard P. *"Surely You're Joking, Mr. Feynman!" Adventures of a Curious Character*. New York: W.W. Norton, 1985. Feynman's engaging collection of tales of his life.

Feynman, Richard P. *What Do You Care What "Other" People Think? Further Adventures of a Curious Character*. New York:

W.W. Norton, 1988. Tales from Feynman's last years, including his service on the *Challenger* disaster investigation committee.
Gleick, James. *Genius: The Life and Science of Richard Feynman.*
New York: Vintage Books, 1992. A good biography of Feynman.
Mehra, Jagdish. *The Beat of a Different Drum: The Life and Science
of Richard Feynman.* Oxford: Clarendon Press, 1994. Biography
with much scientific detail for more advanced readers.

NOTES

p. 84   "You can know the name of that bird . . ." Quoted in Feynman, Richard.
           *"What Do You Care What Other People Think?": Further Adventures of
           a Curious Character.* New York: W.W. Norton, 1988, pp. 13–14.

p. 85   "I thought that was the best . . ." Quoted in Mehra, Jagdish. *The
           Beat of a Different Drum: The Life and Science of Richard Feynman.*
           Oxford: Clarendon Press, 1994, p. 49.

p. 86   "We were sent . . ." Quoted in Mehra, p. 48.

p. 89   "because I like puzzles . . ." Quoted in Gleick, James. *Genius: The Life
           and Science of Richard Feynman.* New York: Vintage Books, 1992, p. 188.

p. 91   "And then, without a sound . . ." Quoted in Gleick, p. 154.

p. 91   "It blasted . . ." Quoted in Gleick, p. 154.

p. 91   "We jumped up and down . . ." Quoted in Gleick, pp. 155–156.

p. 91   "Now I am become death . . ." Quoted in Gleick, p. 156.

p. 91   "Most was known . . ." Quoted in Gleick, p. 204.

p. 96   "What I am really trying . . ." Quoted in Gleick, p. 244.

p. 97   "In private . . ." Quoted in Mehra, p. xxv.

p. 98   "I took this rubber . . ." Quoted in Feynman, pp. 151–53.

p. 98   "The public saw . . ." Quoted in Gleick, p. 424.

6

Hunting the Quark

MURRAY GELL-MANN AND THE EIGHTFOLD WAY

Murray Gell-Mann and other researchers developed the "Eight-fold Way" theory, grouping particles into related "families," *and then discovered the quark.* (Photograph by J. R. Everman, Archives, California Institute of Technology)

By the late 1940s, physics seemed to be settling down again. The inventory of known particles looked like this:

- The **proton**, **neutron**, **electron**, and their anti-particles
- The **photon**, which carried the electromagnetic force
- The **neutrino** and **anti-neutrino**, which were associated with the weak force (nuclear decay)
- Three **pi-mesons** (pions): positive, negative, and neutral, which were associated with the strong nuclear binding force
- Two **mu-mesons** (muons): positive and negative
- **Leptons** (particles like electrons) that are governed by the weak force and not affected by the strong force

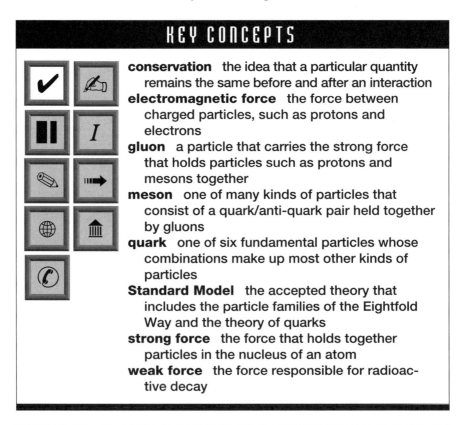

KEY CONCEPTS

conservation the idea that a particular quantity remains the same before and after an interaction

electromagnetic force the force between charged particles, such as protons and electrons

gluon a particle that carries the strong force that holds particles such as protons and mesons together

meson one of many kinds of particles that consist of a quark/anti-quark pair held together by gluons

quark one of six fundamental particles whose combinations make up most other kinds of particles

Standard Model the accepted theory that includes the particle families of the Eightfold Way and the theory of quarks

strong force the force that holds together particles in the nucleus of an atom

weak force the force responsible for radioactive decay

While the role of the muons was still not well understood, in general, the three forces at work in the atom seemed to be accounted for. These were:

- The **strong force**, which bound together the protons and neutrons in the nucleus
- The **electromagnetic force** between charged particles (such as the protons in the nucleus and the circling electrons)
- The **weak force**, which was involved in radioactive decay

The Nuclear Zoo

A stranger then arrived from space. It first announced itself in a V-shaped pair of tracks where a cosmic ray had entered a cloud chamber. An unknown neutral particle, with a mass between that of a proton and a pion, had apparently disintegrated into a pair of pions of opposite charges. Researchers began to search thousands of cloud chamber photographs. They soon found evidence of several more of these "V" particles.

Early in the 1950s, new, more powerful particle accelerators (such as the billion-electron-volt Cosmotron at the Brookhaven National Laboratory in New York) came online. Researchers were now able to create high-energy collisions without waiting for cosmic rays. As a result, several new families of mesons were found. Researchers called some of them "strange particles" because they did not fit into existing theories. The particles also lasted longer than predicted by theory, existing for about a ten-billionth of a second, which in the dizzying new world of high-energy physics was actually considered to be a fairly *long* time.

The floodgates then opened. Led by Enrico Fermi, physicists began analyzing what happens to energy in the collisions that create short-lived particles. They found certain energy levels that made it most likely that a beam of pions in an accelerator would interact with the target protons. These energy levels were called resonances. This is something like finding the pitch or frequency

The first stage of the "tevatron" collider at Fermilab is a linear accelerator. It is like a giant power transformer that uses a high voltage to accelerate charged particles and inject them into the tunnel ring. (Fermilab Visual Media Services)

of sound that makes a string in a piano start to vibrate in sympathy. The energy turned into mass in the collision, and in fact acted like a particle with characteristics such as mass and spin. Thus, although these resonance "particles" could not be detected directly, they were "entitled" to be treated like other particles.

TRENDS

ACCELERATORS: PHYSICS' HEAVY ARTILLERY

The following table gives examples of important particle accelerators. The power is the collision energy in MeV (millions of electron volts). Note that some of these machines have been modified to work in new ways (such as by becoming colliders where beams of particles slam into one another).

DATE BUILT	NAME	LOCATION	TYPE	POWER (MeV)
1928	Walton-Cockroft	Cambridge, England	Transformer-rectifier	0.75
1952	Cosmotron	Brookhaven National Laboratory, Long Island, New York	Proton synchrotron	3,000
1954	Bevatron	University of California, Berkeley	Proton synchrotron	6,400
1960		CERN, Switzerland	Proton synchrotron	28,000
1972		Fermilab, Chicago, Illinois	Proton synchrotron	400,000
1972	SLAC	Stanford University, California	Linear accelerator (positron-electron)	100,000
1984	Tevatron	Fermilab, Chicago, Illinois	Proton synchrotron	1,000,000

Workers doing maintenance in the main tunnel at the Fermilab tevatron
(Fermilab Visual Media Services)

During the 1960s and 1970s, hundreds of such "particles" were found. One textbook of the time included a list of "particles we could do without." It was as though a naturalist had turned over a rock and found endless numbers of insects that fit into no known species. Physicists were challenged to find characteristics that they could use to organize the growing number of particles and make sense of their interactions.

The Language of Science

Perhaps it was appropriate that the physicist who would do the most to discover the "natural history" of atomic particles had an older brother who taught him to love bird-watching as well as science and literature. Murray Gell-Mann (1929–), like Richard

Feynman, grew up in New York and became a world-class nuclear physicist.

Murray's path to a scientific career was less direct than Feynman's. His father ran a language school for immigrants, and the study of languages would become a lifelong interest. Murray's brightness was recognized early, and he was transferred to a school for gifted children. He found, however, that school was "dull, dull, stuff" and that "physics in high school was terribly boring."

At the age of only 15, Murray was accepted for Yale University. Like many precocious youngsters who find themselves in classes where most everyone is several years older, Gell-Mann had some trouble adjusting socially. He did not seem to have the strong pull toward physics or mathematics of most of the physicists featured in this book. He later said that becoming a physicist "just happened."

In 1948, Gell-Mann graduated from Yale and went to the Massachusetts Institute of Technology—the same school that had somewhat disappointed Feynman about ten years earlier. He was fortunate to find an adviser, Victor Weisskopf, who gave him the first real taste of what it was like to work as a professional physicist. He also challenged young Gell-Mann to tackle interesting problems.

"Strange" Particles

After receiving his doctoral degree in 1951, Gell-Mann got a teaching job at the University of Chicago, where researchers working under Enrico Fermi were trying to make sense of the dozens of new particles that were being detected in experiments.

Many of these particles were born in violent atomic collisions and lived for only a few millionths of a second before decaying. Still, existing theory did not expect them to live even that long. The particles were manifestations of the strong force within the nucleus, but when they decayed, they did so because of the weak force.

Gell-Mann decided to look for a characteristic that would account for this behavior. He found that he could assign a new quantum number that he called "strangeness" to the particles in such a way that the equations that described their interactions "balanced" according to all the accepted rules.

Gell-Mann and other researchers were helped in their effort to categorize the new particles by the fact that particles obey "conservation laws." These are laws that state that the total amount of something (such as mass-energy, momentum, spin, and electric charge) at the end of an interaction must be the same as the amount at the beginning. This is something like balancing the books of a business where credits and debits must add up to the same total.

Conservation had several benefits for physicists. First, it ruled out many possible results from particle interactions because they broke one or more of the conservation rules. Even more surprising, it was found that any particle that was not *forbidden* by a conservation rule from decaying would decay, and do so as soon as possible! These two principles were very helpful in looking at the expected behavior of particles. By using conservation and his new "strangeness numbers," Gell-Mann noted:

> I was able to predict a number of strange particles. I predicted that experimental physicists would find certain ones; that they would *not* find others. And this was all true. They actually found all those that I predicted and didn't find any others.

(A Japanese physicist, Kazuhiko Nishijima, working independently of Gell-Mann, developed a similar theory.)

The Eightfold Way

In 1955, Gell-Mann went to the California Institute of Technology (Caltech), where he and Richard Feynman would work together until the latter's death in 1988. The two men certainly clashed at times, since their personalities and styles were so different. Although Gell-Mann was about ten years younger, it was Feynman who often seemed to have the attitude of a rebellious youth in fighting with the bureaucracy. Gell-Mann wore a tweed suit; Feynman went around in shirtsleeves. Sidney Coleman, one of their colleagues, said, "Murray's mask was a man of great culture. Dick's mask was Mr. Natural—just a little boy from the country that could see through things the city slickers can't."

In his scientific work, Gell-Mann insisted on a precisely constructed mathematical proof, even if it would be hard to explain. Feynman, on the other hand, was willing to try anything that worked and plug up the holes later. He believed, perhaps in the spirit of the uncertainty principle, that theories themselves would always be only approximately true. But such a contrast is overly simple and only somewhat accurate. Gell-Mann could be creative, Feynman meticulous.

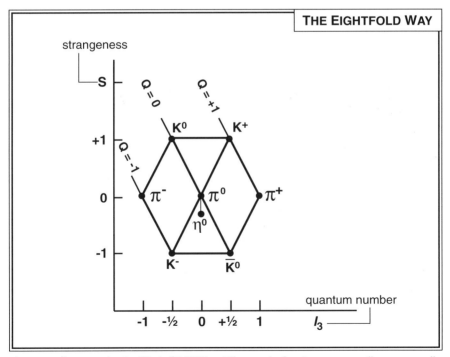

A group of mesons in the Eightfold Way. The vertical axis represents "strangeness" and the horizontal axis another quantum number.

Tools from abstract mathematics have had a way of popping up and determining the direction of physics research. During the 1920s, Werner Heisenberg and his followers had used number groups called matrices to manipulate the numbers that described the electron. As the 1960s began, Gell-Mann took a related branch of mathematics, the theory of groups, and applied it to the quantum numbers of the newly discovered particles. He found that he could group particles in "families" of differing sizes.

For example, one can take all the mesons that have zero spin and negative parity (parity is discussed in Reaching for Unity). By plotting these eight particles on a graph that has the strangeness number on the vertical axis and another quantum number called I_3 along the horizontal axis, the particles fall into a hexagonal arrangement. Many other particle groups fell into similar arrangements. Gell-Mann (and the developer of a similar theory, the Israeli physicist Yuval Ne'eman) concluded that the particles in each group could be considered to be different "states" of the same underlying particle, in somewhat like the way that ice, liquid, and vapor are all states or forms of water.

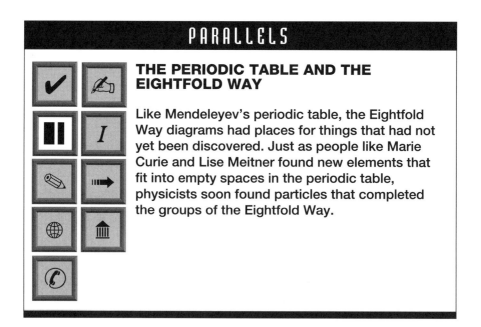

PARALLELS

THE PERIODIC TABLE AND THE EIGHTFOLD WAY

Like Mendeleyev's periodic table, the Eightfold Way diagrams had places for things that had not yet been discovered. Just as people like Marie Curie and Lise Meitner found new elements that fit into empty spaces in the periodic table, physicists soon found particles that completed the groups of the Eightfold Way.

The researchers then found a new quantum number that could be used to combine the groups into larger "super-groups." Because they used eight combinations of quantum numbers to create the entire system, Gell-Mann named it the "Eightfold Way." This was a reference to the eight commands that Buddha had given for people who want to live a righteous life.

Between 1961 and 1964, the remaining members of three of the groups were found. The greatest triumph was the finding of a predicted particle called Omega Minus that completed a group of twelve. It took an examination of 100,000 photographs from a liquid hydrogen bubble chamber to reveal an unmistakable trace of the elusive particle. This result, according to the director of the Brookhaven National Laboratory, "formed the capstone of a building which was held together only by the bold imagination of Dr. Gell-Mann and Dr. Ne'eman." Gell-Mann won the Nobel Prize in physics in 1969.

Quarks

The Eightfold Way had revealed a useful web of relationships—a kind of family tree of particles. But just as the discovery of DNA was needed in order to finally learn how living things inherit characteristics, particle physicists still needed a theory that could show how the particle families came into being.

In 1964, Murray Gell-Mann and George Zweig worked independently to come up with a theory that built all atomic particles from combinations of a small number of components. While Zweig called these ultimate particles *aces*, it was Gell-Mann's word that stuck: *quarks*. (The linguistically adept Gell-Mann later said he had the sound in mind and then came up with the word, which is found in the novel *Finnegans Wake* by James Joyce.)

The theory began with three quarks called *u*, *d*, and *s* for up (upward spin), down (downward spin), and "strangeness." Like regular particles, each quark also had an "anti-quark" twin. Each meson consists of a pair of a quark and an anti-quark. Each baryon

CONNECTIONS

ELEVEN BILLION DICE

Quark theory had shown that pairs or triplets of quarks combined to form most atomic particles. But exactly how did this process take place? What laws governed the behavior of quarks?

Finding the answers would be a formidable task. In addition to the six types of quarks there are three characteristics called "colors," arbitrarily labeled red, green, and blue. Since every quark had an anti-quark, that meant 36 combinations in all.

The method of calculating the possible values of the "chromoelectric field" of quarks and their color "charges" uses a lattice, a grid with four dimensions (the regular three space dimensions plus time). The number of calculations needed was astronomical—about equal to a one followed by 96,000 zeroes. Researcher Donald Weingarten and his colleagues came up with some shortcuts, such as a way to sample probabilities ("roll the dice") rather than doing all the calculations. Even using shortcuts, though, they estimated that it would take a hundred years to calculate the steps leading to the formation of a single particle.

In 1983, however, Weingarten and his team used a new idea called parallel processing and began to design a different kind of computer. They combined many separate computer processors that each calculated a portion of the result. Their computer, called GF11, performed 11 billion calculations per second. With it they were able to predict a previously unknown particle in 1995.

The computer field has progressed so fast, however, that today about 200 desktop computers with Pentium chips could do the same job!

(particles like protons and neutrons) consists of three quarks, and the corresponding anti-particle consists of three anti-quarks. Thus a proton is made up of two up quarks and a down quark, while a neutron has two down quarks and one up quark.

A few more kinds of quarks proved to be necessary for the complete theory. The "charm" quark helped explain the relationship between leptons (particles like the electron that are not made of quarks) and the quarks. Two more quarks, called "top" and "bottom" (some used the names "truth" and "beauty" at first) were found to account for superheavy particles found in high-energy experiments in the 1970s and 1980s. The top quark, most massive of all, was the hardest to find. In 1994, a computer-aided search went through one trillion particle collisions to find 20,000 that involved a W boson—a particle that results from the decay of a top quark. Eventually 12 events were found that showed three ways that top quarks can decay into other particles.

The behavior of quarks was explained by a theory called quantum chromodynamics, where the *chromo* means "color," an arbitrary name for a quantum characteristic. This characteristic causes the strong nuclear binding force, which is passed through particles called gluons. The binding force binds together the quarks into combinations that make up most kinds of particles.

A further refinement of the theory groups quarks into "generations." Each generation consists of one pair of quarks and one pair of leptons. The first generation contains the down and up quarks, the electron, and the electron-neutrino. The second generation has the strange and charm quarks and the muon and muon-neutrino. Finally, the third generation brings the bottom and top quarks and the tau and tau-neutrino. But were there any more generations? Surprisingly, physicists, with the aid of new superpowerful colliders built in the 1980s, were able to show that there were only three.

In recent years Gell-Mann has sought to relate the discoveries of quantum physics to the ways living things, including humans and their cultures, change and adapt. He notes:

Worldwide headlines announced the discovery of the "top" quark. (Fermilab Visual Media Services)

ISSUES

LEVELS OF COMPLEXITY

In a seminar where ten Nobel Prize winners were asked to discuss issues relating to science, culture, and education in the future, Murray Gell-Mann was asked whether, as the new millennium approached, scientific thinking was moving from "reductionism" (finding a few simple underlying rules to explain complex phenomenon) to "holism" (seeing how everything is connected to everything else in an "organic" kind of way). Gell-Mann's answer suggests that there should be no conflict between these kinds of thinking:

One point on which I most emphatically disagree with this question has to do with the fate of the idea of reduction. It is absolutely unnecessary to discard the idea of reduction, in order to accomplish the other things. . . . For example, chemistry can be reduced to physics, to quantum mechanics and electromagnetism plus the accidents of history that create conditions of temperature and pressure that allow atoms and molecules to exist. . . . This capacity for reduction in principle does not in any way detract from the importance of studying chemistry at its own level, in terms of chemical bonds, chemical reactions, chemical species and so on. . . .

Biology emerges from physics and chemistry plus the unpredictable accidents of the evolution of life here on earth. Human consciousness, I am

(continued)

ISSUES

(continued)

sure, is no different, and emerges in the same way from the laws of biology plus certain accidents of primate evolution. But that does not mean that biology should not be studied and respected for its own sake, at its own level, with its own laws and its own interesting peculiarities and that human nature should not receive special attention as an interesting subject in its own case. And when human beings exhibit love for one another and care about one another or about the organisms with which we share the biosphere, when human beings create great works of art or of the intellect, these wonderful developments lose none of their wonder through being in principle reducible to a more primitive, to a more fundamental level.

. . . since 1924, we have learned that there is a huge fundamental source of indeterminacy, namely quantum mechanics. Put all of these together and you see the world is subject to an enormous amount of indeterminacy, and this is where complexity arises. Accidents occur. Some of these are what we call frozen, which can have massive ramifications through history or through huge regions of space and time and the accumulation of those accidents makes complexity possible.

Complex adaptive systems arise to take advantage of these accidents. The best known, of course, is life itself. On Earth, biological life has exploited a planet, at the right temperature, with the right raw materials at hand, to evolve into a myriad of niches, and to continue evolving as the environment changes and new opportunities arise.

In studying "complex adaptive systems," Gell-Mann has brought together his lifelong interests in physics, nature, and language. Meanwhile, physics struggles toward a "theory of everything" that could explain not only particles but the origin of the four basic forces (the nuclear forces plus gravity) and the universe itself.

Chronology of Murray Gell-Mann and the Eightfold Way

1953	Gell-Mann writes paper explaining concept of "strangeness" in particles
1961	Murray Gell-Mann and Yuval Ne'eman discover the Eightfold Way patterns in "particle families"
1964	Murray Gell-Mann and George Zweig propose the existence of quarks as fundamental particles
1969	Gell-Mann wins Nobel Prize in physics for quark theories
1970	Sheldon Glashow, John Iliopoulos, and Luciano Maiani propose the charm quark
1974	Discovery of psi meson by Burton Richter and Samuel Ting confirms existence of charm quark
1977	S. W. Herb finds the upsilon resonance implying the existence of the bottom (or "beauty") quark
1989	Theory of three quark "generations" confirmed
1995	Conclusive evidence found for "top" quark

Further Reading

Berland, Theodore. *The Scientific Life*. 1962. Contains an interview with Murray Gell-Mann.

Current Biography Yearbook. New York: H.W. Wilson Co., 1940–1966 annual. "Murray Gell-Mann." Article summarizing Gell-Mann's career up to his discovery of the Eightfold Way.

Gottfried, Ted. *Enrico Fermi: Pioneer of the Atomic Age*. New York: Facts On File, 1992. Young adult biography.

Spangenburg, Ray and Moser, Diane K. *The History of Science from 1946 to the 1990s*. New York: Facts On File, 1994. Young adult book that contains a section on Gell-Mann's work with quarks.

Ten Nobels for the Future: Science, Economics, Ethics for the Coming Century. A conference held in Milan, Italy, 5–7 December 1995. Selected answers to questions to Nobel Prize winners can be found on the World Wide Web at www.smau.it/nobel/nobel95/direct/direct.htm.

Weingarten, Donald H. "Quarks by Computer." *Scientific American*, February 1996, pp. 116–20. Describes the use of computers for solving problems in quark physics.

NOTES

p. 108   "I was able to predict . . ." Quoted in *Current Biography Yearbook*. New York: H.W. Wilson Co., 1940–1966 annual, p. 125.

p. 109   "Murray's mask . . ." Quoted in Gleick, p. 389.

p. 111   "formed the capstone . . ." *Current Biography Yearbook*, p. 126.

pp. 115–16 "One point on which . . ." Quoted in *Ten Nobels for the Future: Science, Economics, Ethics for the Coming Century*. World Wide Web: www.smau.it/nobel/nobel95/direct/direct.htm.

p. 116   ". . . since 1924 we have . . ." Quoted in Coyne, Pat. "Quark songs." Interview with Murray Gell-Mann. *New Statesman & Society*. July 1, 1994, p. 29.

Reaching for Unity

PHYSICS AT THE END OF THE TWENTIETH CENTURY

With the proof of the existence of quarks, what has become known as the Standard Model of particle physics seems to be well established. But just as physics at the end of the 19th century turned out to be incomplete, there are still important questions that physicists are working on today. These questions include

- Are quarks the most fundamental particles, or is there something simpler?
- Is there any limit to the number of different kinds of particles?
- Do the three fundamental atomic forces plus a fourth force, gravity, have a common origin?
- How is the origin of the universe related to the laws of physics we see today?

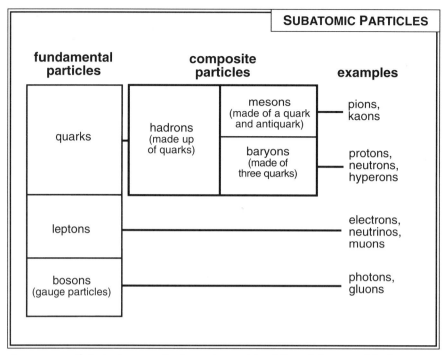

A summary of the basic types of subatomic particles. Most of these particles have corresponding anti-particles.

Are Quarks Really Elementary?

The Standard Model suggests that quarks are "fundamental particles"—that is, they cannot be broken down into smaller pieces. But using an accelerator to smash protons and anti-protons together at nearly two *trillion* electron-volt energies, researchers in the 1990s have been able to observe "jets" or streams of high-energy particles coming out of the collision in the same direction. Some of these jets appear to have much more energy than predicted by theory. While the problem may be in the measurement itself, it is possible that the quarks themselves have massive particles hidden within them that are participating in the collisions. These possible particles have been dubbed "preons," and a variety of theories have attempted to show how the "generations" of quarks and leptons might be constructed from them.

Why So Many Particles?

Whether or not quarks turn out to have other particles hiding in them, there's also the question of why there seem to be so many "fundamental" particles. As Murray Gell-Mann points out, there's something unsatisfactory about the Standard Model still having about 60 different particles: quarks, leptons, gluons, and bosons. Gell-Mann agrees that "to a lay observer, it seems crazy to suppose that the basic law of all matter in the universe could rest on such a large and heterogeneous [unlike] collection of fundamental particles."

What are the possibilities? There could be some final level, perhaps below that of quarks, at which there are only a relatively few kinds of particles. Gell-Mann believes, however, that this is unlikely because explaining the variety of characteristics found in particles seems to require a certain minimum number of "fundamental" particles.

Another possibility is expressed in the story of the lady who told a scientist that she believed that the earth was carried on the back

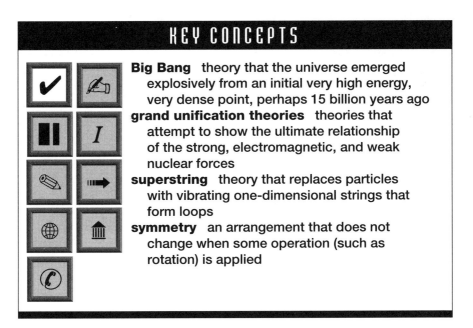

KEY CONCEPTS

Big Bang theory that the universe emerged explosively from an initial very high energy, very dense point, perhaps 15 billion years ago

grand unification theories theories that attempt to show the ultimate relationship of the strong, electromagnetic, and weak nuclear forces

superstring theory that replaces particles with vibrating one-dimensional strings that form loops

symmetry an arrangement that does not change when some operation (such as rotation) is applied

of four turtles. When the scientist asked her what the turtles stood on, she said "some larger turtles." When asked what those larger turtles in turn stood on, she called a halt to the questioning by insisting, "It's turtles all the way down." In other words, there might be no limit to how many layers of particles might be found: quarks, preons, something-else-ons . . .

A third possibility, according to Gell-Mann, is:

> that a simple theory underlies the elementary particle system, according to which the number of such particles can be regarded as infinite, with only a finite number accessible to experimental detection at available energies.

Superstrings

One theory that yields infinite possible particles is called superstring theory. According to this theory the smallest things in the universe are not tiny, almost pointlike particles, but rather one-dimensional vibrating "strings" that can twist and turn in various ways. The theory uses a mathematics with a mind-boggling ten dimensions rather than the four (length, width, height, and time) that we can experience in our ordinary world.

A version of superstring theory introduced in 1984 seemed to incorporate both the Standard Model of quantum theory and gravity. Gravity is much weaker than the nuclear forces, but it is most important once one gets beyond the tiny distances within the atom. While a theoretical particle called a graviton has been proposed for carrying gravitational force, no one has yet been able to come up with an experiment that can detect it.

When the superstring theorists tried to get from ten dimensions down to the ordinary four, however, the equations seemed to yield an infinite number of possible solutions. If the theory is to lead to a true "theory of everything," it must somehow be shown how it can get to the particular solution that represents the actual universe we see.

The Broken Mirror

Another angle for seeing the new physics comes from the effort to show how the three nuclear forces (and eventually, gravity) might be aspects of a single interaction. This involves the concept of symmetry.

Symmetry is closely related to conservation (see Hunting the Quark). A symmetry can be defined as something that does not change when a certain operation is applied. For example, integer numbers are "symmetrical" with respect to addition: it doesn't matter whether one adds 5 + 7 or 7 + 5.

Nature is full of symmetry. Our bodies are roughly symmetrical: if you draw a line down from the top of your head you will find that eyes, ears, lungs, and kidneys, not to mention arms and legs, are neatly paired on either side of the line. If you look at yourself in a mirror, things are reversed left to right, but the pattern remains the same.

In physics one of the most important kinds of symmetry involves parity, or the lack of "handedness." Physicists thought for many years that the universe did not really care whether things went from left to right or right to left. Any particle interaction should be "flippable" left for right as though viewed in a mirror.

In 1956, Richard Feynman and Martin Block said, in effect, "Wait a minute! Are we *sure*?" Two Chinese-American physicists, Tsung-Dao Lee and Chen Ning Yang, surveyed the known kinds of particle interactions. The strong nuclear interaction and the electromagnetic interaction seemed to definitely have parity. But what about the weak interaction of radioactive decay? There, they were not so sure. They proposed experiments to find out.

A few months later, another Chinese-American physicist, this time a woman named Chien-shiung Wu, designed an experiment that cooled the radioactive isotope cobalt 60 close to absolute zero and turned on a powerful magnet. The magnetism essentially aligned the cobalt atoms in the same direction so the experimenters could compare the directions in which beta particles shot out from them as they decayed. Wu determined that the normal and mirror image were not the same, after all—in the normal image, most of the beta particles went off in a direction opposite to the magnet effect, while in the mirror image they went in the same direction.

As physicists looked at whether each of the various kinds of symmetry and conservation held true for the strong, or electromagnetic, or weak force, a pattern emerged. The strong force had the most kinds of conservation, the electromagnetic force had almost as many, while the weak force had only about half as many. The strong force affected the fewest number of particles but had the most symmetry, while the weak force affected more particles but had less symmetry. This suggests that perhaps the universe started out having only the strong force (and full symmetry), but that the symmetry "broke down" as the energy level decreased, bringing the other forces into play and allowing for more kinds of particles. The key to a unified theory, therefore, might lie in the earliest moments of the universe when symmetry had not yet been broken.

Back to the Beginning

Physicists have been working toward a "grand unification theory" or GUT, that if successful would show the strong, electromagnetic, and weak forces to be all manifestations of the same underlying force. Similarly, the difference between quarks and leptons would vanish at some point. The problem is that this convergence of forces and particles is believed to happen only at energies perhaps a trillion times that which can be reached by today's particle accelerators.

According to cosmologists (astrophysicists who study the beginnings of the universe), the universe we know today began perhaps 15 billion years ago in an explosion called the Big Bang. In the first tiny fractions of a second the incredibly dense, hot "stuff" expanded and cooled to begin to form the matter that later became galaxies, stars, and planets.

ISSUES

HAS "BIG PHYSICS" REACHED ITS LIMIT?

It was to be called the Superconducting Supercollider, or SSC. Consisting of an oval ring 54 miles (87 km) in circumference, it would have accelerated and collided particles at energies of up to 40 TeV ("tera" or trillion electron volts), more than 20 times more powerful than any existing accelerator. In June 1993, however, the U.S. House of Representatives voted down further funding for the $10 billion project amid charges of mismanagement.

Some physicists had complained that this huge expenditure would take away funding from other fields of physics. Other critics opposed spending such a large amount of money to just satisfy the curiosity of theoretical physicists.

The criticism of "big physics" has been similar to that of the expensive NASA space missions and space station. NASA has recently changed its emphasis to flying a larger number of smaller,

(continued)

The main control room at Fermilab. More powerful accelerators may reveal evidence for unification theories, but their cost makes them controversial. (Fermilab Visual Media Services)

ISSUES

(continued)

cheaper, less ambitious missions. It is not clear how a similar approach in physics might work.

Another possibility, in both space and physics, is to have more international projects that pool resources from several countries. Existing particle accelerators can also be upgraded to some extent as new technology becomes available.

Today's astronomical instruments (such as the orbiting Hubble Space Telescope) can study objects billions of light-years away. Since the light or other radiation we detect left such objects billions of years ago, it tells us about the conditions in the universe closer and closer to the original Big Bang. Perhaps as they go back in time, researchers will be able to observe the conditions shortly after the Big Bang before the different nuclear forces and particles became separated.

This book opened with the end of the 19th century when scientists thought physics had reached a final conclusion. Today, on the threshold of the 21st century, how close is a final "theory of everything" that can explain the world within the atom? Let the physicists have the last word:

Stephen Hawking, the famous physicist who overcame a severe physical handicap to uncover the mystery of black holes, seems to be cautiously optimistic:

> If we do discover a complete theory, it should in time be understandable in broad principle by everyone, not just a few scientists. Then we shall all, philosophers, scientists, and just ordinary people, be able to take part in the discussion of why it is that we and the universe exist. If we find the answer to that, it would be the ultimate triumph of reason—for then we would know the mind of God.

But Richard Feynman, in the tradition of uncertainty at the heart of quantum physics, expresses a different attitude:

> People say to me, "Are you looking for the ultimate laws of physics?" No, I'm not . . . If it turns out there is an ultimate law which explains everything, so be it—it would be very nice to discover. If it turns out it's like an onion with millions of layers . . . then that's the way it is . . . .

The asking of new questions and the search for answers continue.

Chronology of Recent Discoveries in Nuclear Physics

1967   Steven Weinberg proposes theory unifying electromagnetism and weak nuclear force

1972   Stephen Hawking, James Bardeen, and Brandon Carter explain physics of black holes

1974   Stephen Hawking applies quantum theory to black holes

1981   "Superstring" theory introduced

1984   String-based "theory of everything" proposed

1993   U.S. Congress cancels funding for Superconducting Supercollider

Further Reading

Cramer, John G. "Inside the Quark." *Analog Science Fiction and Fact*, September 1996, pp. 91–95. Summarizes interesting recent research that suggests quarks may be made of smaller particles.

Gell-Mann, Murray. *The Quark and the Jaguar: Adventures in the Simple and the Complex*. New York: W.H. Freeman, 1994. Describes Gell-Mann's work with complexity in physics and in life; includes a good introduction to the Standard Model and the new superstring theory.

Henderson, Harry. *The Importance of Stephen Hawking*. San Diego, Calif.: Lucent Books, 1995. Biography of Stephen Hawking for young people; includes introduction to black hole physics and cosmology.

NOTES

p. 121   "to a lay observer . . ." Quoted in Gell-Mann. *The Quark and the Jaguar: Adventures in the Simple and the Complex*, New York: W. H. Freeman, 1994, p. 197.

p. 122   "that a simple theory . . ." Quoted in Gell-Mann, p. 198.

p. 126   "If we do discover . . ." Quoted in Hawking, Stephen. *A Brief History of Time*. New York: Bantam Books, 1988, p. 175.

p. 127   "People say to me . . ." Quoted in Gleick, p. 432.

Index

. . . .

Page numbers in *italics* indicate illustrations. Page numbers in **boldface** indicate major treatment of topics.